MILTON D. HEIFETZ / WIL TIRION

A WALK
THROUGH THE
HEAVENS

Third edition

A GUIDE TO STARS

AND CONSTELLATIONS

AND THEIR LEGENDS

CAMBRIDGE
UNIVERSITY PRESS

PUBLISHED BY THE PRESS SYNDICATE OF THE UNIVERSITY OF CAMBRIDGE
The Pitt Building, Trumpington Street, Cambridge CB2 1RP, United Kingdom

CAMBRIDGE UNIVERSITY PRESS
The Edinburgh Building, Cambridge CB2 2RU, UK
40 West 20th Street, New York, NY 10011-4211, USA
477 Williamstown Road, Port Melbourne, VIC 3207, Australia
Ruiz de Alarcón 13, 28014 Madrid, Spain
Dock House, The Waterfront, Cape Town 8001, South Africa

© Milton Heifetz, maps Wil Tirion 1996, 2004

First published 1996
Second edition 1998
Third edition 2004

Printed in Great Britain at the University Press, Cambridge

Typeset in ITC Century Light 10/13pts

A catalogue record for this book is available from the British Library

ISBN 0 521 544157 paperback

Illustrations by Wil Tirion

Book design and typesetting by Kevin McGeoghegan MCSD

Contents

This book is dedicated to our grandchildren

Elena, Sari, Ariel, Jenny, Litan, Ilan, David, Ariana, Ori, and those yet to come

Acknowledgements

I wish to acknowledge the excellent suggestions I have received from Larry Schindler of the Smithsonian—Hayden Planetarium of Boston and the late George Lovi of the Hayden Planetarium of New York. Their meticulous critique of the manuscript is greatly appreciated. I also wish to express my gratitude to Robert Kimberk and Freeman Deutsch of the Harvard University Center of Astrophysics, Van Del Chamberlain of the Hansen Planetarium in Salt Lake City and to William Luzader.

Introduction

This book is written for those who look at the stars with wonderment and would like to feel more at home with them, to go for a friendly walk with them.

In order to walk through the heavens and to know where you are, you must recognize what your eye sees. To call your neighbor by his or her name is the beginning of friendship. To know the names of stars and constellations is to form a friendship with our heavenly neighbors.

As we walk among the constellations, you will feel the immensity and quiet peace of the night sky. Do not ignore the legends about the constellations in Part 3 of the book. These legends will lend greater feeling to your vision of the world above. Friendship with the stars will deepen as we sense the thoughts and dreams of people who imagined people and animals living among the constellations.

Our walk will take us to the brightest stars in the sky. When we become familiar with them they will lead us to the dim stars.

It is not enough simply to find a constellation. Try to see relationships between constellations. This is best done if you know different pathways to the constellations.

Almost certainly the early cave dwelling people and probably the early forms of humans looked to the stars with a sense of awe. In their vivid imagination they joined certain bright stars together into patterns in which they imagined figures of animals or people with unusual attributes. These patterns, that we now call constellations, varied among different peoples across the world.

From the earliest of times people have looked to the stars to help them navigate across seas and deserts, know when to plant and harvest, establish their legends, mark the change of seasons and even to align their temples of worship.

Constellations were recorded over 5000 years ago and lists of such patterns were written 2400 years ago by the Greek astronomer Eudoxus who studied under Plato. Ptolemy, who lived 2100 years ago, compiled a list of 48 constellations which has remained relatively standard to this day. Later, Johann Bayer (1572–1625), Johannes Hevelius (1611–1687) and Nicolas de Lacaille (1713–1762) added more constellations to the list. Professional astronomers now officially recognize 88 constellations which they regard simply as areas of the sky, not as star 'pictures' or patterns. These patterns have never been made 'official', so you should feel free to make any constellation design you wish.

Before we begin our walk through the heavens, we should understand two concepts: how to measure distances in the sky, and the brightness of the stars. After this is done, follow the instructions on how to use the atlas to best advantage.

In Part 2, 'A walk through the heavens', the design or picture of a group of stars to form a constellation image will usually, but not always, contain stars which are bright enough to be seen easily. Most of the constellation patterns are well recognized images, but some are new.

For convenience, each star in each constellation will be numbered and some will be named, so that we can more easily identify specific stars to help us walk around the sky. We will follow several paths to a constellation. By doing this you will have a better sense of star relationships.

Since I have been disturbed with the violence that is part of the commonly used legends associated with the constellations, I have taken the liberty of modifying and abridging them. Legends have been and will continue to be modified with each generation.

This book applies to people living in the Northern Hemisphere.

Relax and enjoy yourself as you travel across the sky.

Part 1 Measuring distances in the sky

How do we measure the size of
the Big Dipper or the distance
between two stars? We cannot
measure these distances in
inches or millimeters, which are
linear measurements
(measurements along a line).
Instead, we must use a
measuring system using angles
to determine how far apart one
star or constellation may be from
another.

Fig.1

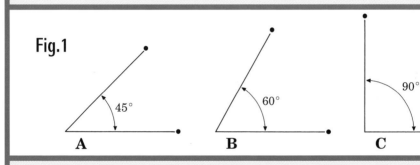

A   B   C

An angle is formed by two lines
that meet at one point. The
space between the two lines
forms an angle that is measured
in **degrees**. There are 360
degrees in a circle. The angle in
Fig. 1A is 45 degrees; 1B is 60
degrees; 1C is 90 degrees. The
further apart two stars, the
larger the number of degrees
between them.

How do we measure angles in
the sky in a practical way
without fancy instruments?

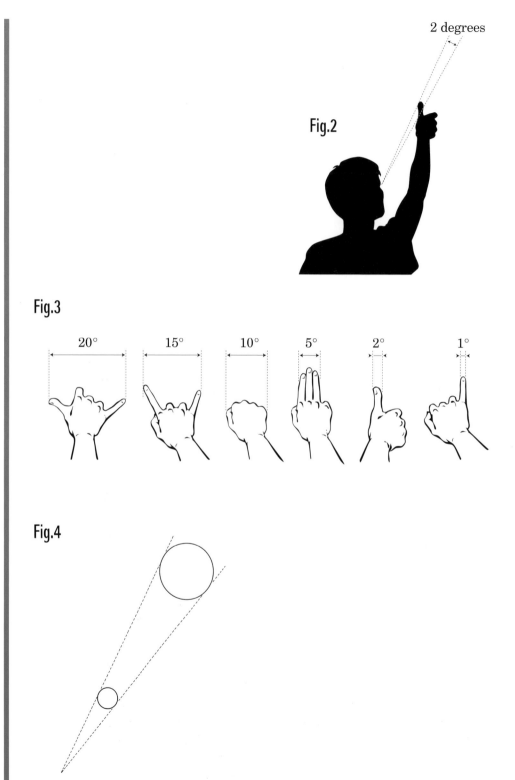

2 degrees

Fig.2

Fig.3

20°   15°   10°   5°   2°   1°

Fig.4

We measure the distance between two far away objects such as stars using our eye as the corner of the angle and using a part of our hand to hide the sky between those stars. The further apart the stars are, the more of our hand we need to use to cover the space between them. Look at Fig. 2.

With your arm outstretched, your hand will help you determine angular distances. Extend your arm out in front of you and hold your thumb upright. It is now hiding part of what is in front of your vision. The amount of view that is hidden behind your thumb will depend upon how long your arm is and how thick your thumb may be. The shorter the arm, or the thicker the thumb, the more of your view will be hidden.

So our hand becomes an excellent device for measuring distances in degrees in the sky. Different parts of your hand can be used to measure different angles. Look at Fig. 3.

The tip of your small finger will cover approximately 1 degree of sky. In your room look at the door knob or light switch across the room. Your finger can cover it. Now look at a building across the street. The same finger will cover a large part of the building. Now look at the Moon. The same finger can cover the Moon. How can this be since one is so much larger than the other? Look at Fig. 4.

Although the Moon is so much larger than the building across the street, it can actually be hidden by a narrow object like a finger. The diameter of the Moon, when measured this way, is seen to be only about 1–2 degree wide. The farther away an object is, the smaller the angle needed to hide it from sight. The Moon looks much bigger than a star because it is so much closer to us.

Distances to the stars
We measure the distance between a star and the Earth, not in miles or kilometers, but in **light years** by using the speed of light. It is important to remember that a light year is a distance; it is not a measure of time. The distance light travels in one year is a light year. Light travels 186 000 miles per second (299 000 kilometers per second), which is 680 760 000 miles per hour (1 096 000 000 kilometers per hour). Therefore, a light year is a distance of almost 6 000 000 000 000 (6 trillion) miles, or 9.6 trillion kilometers.

It takes slightly more than one second for light from the Moon to reach the Earth and more than 8 minutes for light from the Sun to reach Earth. Compare this with the 4.3 years that it takes for the light from the nearest star, Alpha Centauri, to reach the Earth. Deneb in the Northern Cross is over 1000 light years away. That means the light we now see left the star over 1000 years ago. It is therefore possible that the star may not even be there any more.

The brightness of stars
Some stars appear much brighter than others. This does not necessarily mean that the bright star is bigger or giving off more light than the dimmer star. The **apparent brightness** (how bright it seems to us) depends upon three

things: (1) how big it is; (2) how far away it is from Earth; and (3) how much light it actually emanates per diameter of the star. The brightest star to us is our Sun, but it is only an average size star – it seems the brightest because it is the nearest star to us on Earth.

The star Sirius in the constellation of Canis Major appears considerably brighter than Rigel in Orion. However, Rigel is actually thousands of times brighter than Sirius. It appears fainter because it is over a thousand light years away, while Sirius is only 81–2 light years from us.

We measure the brightness of the stars as seen with the naked eye on a scale called the **magnitude** scale. Hipparchus, a Greek astronomer, rated the importance of stars by their brightness and used the word magnitude to describe their relative brightness. Magnitude means bigness. In ancient times they may have assumed that the brighter star is a bigger star. A very bright star would have a magnitude of 1 or less and a very faint star a magnitude of 6. The smaller the number, the brighter the star. A very powerful telescope can see very faint stars beyond magnitude 20. You may be able to see stars with a magnitude of 6 to 7 with your naked eye under very clear, moonless skies. The very brightest planets have a magnitude of –1 to –4. Unfortunately, light pollution from home and street lamps may prevent you from seeing as many stars as you could if your surroundings were in total darkness.

Remember, magnitude measures how bright the star is to our naked eye, not how much light the star actually produces, nor how big it is.

Although there are billions of stars, we can only see approximately 2500 stars with our naked eye at one time under the best of conditions. Later look on page 67 and read how to test your vision.

The Milky Way

The space around us seems to be endless. It is a space occupied by billions upon billions of galaxies, each of which is composed of billions and billions of stars, of which our Sun is an average-sized example. The faint band of stars that arches across the sky was called the Milky Way by the early Greeks. It is our view of the galaxy in which we live from within one of our galaxy's spiral arms. The location of our Sun and Earth in that spiral arm is approximately 30 000 light years from the center of our galaxy.

What we see with our naked eye is confined to our own galaxy. However, with good eyesight, and if the night is dark enough, you may see a neighboring galaxy as a faint blur in the constellation Andromeda, or the Small and Large Magellanic Clouds in the region of the constellation Hydrus. Although our galaxy is whirling in space at tremendous speed it still takes 225 million years to complete one revolution. That time period is called the Galactic Year.

Fig. 5 – Side view of a galaxy

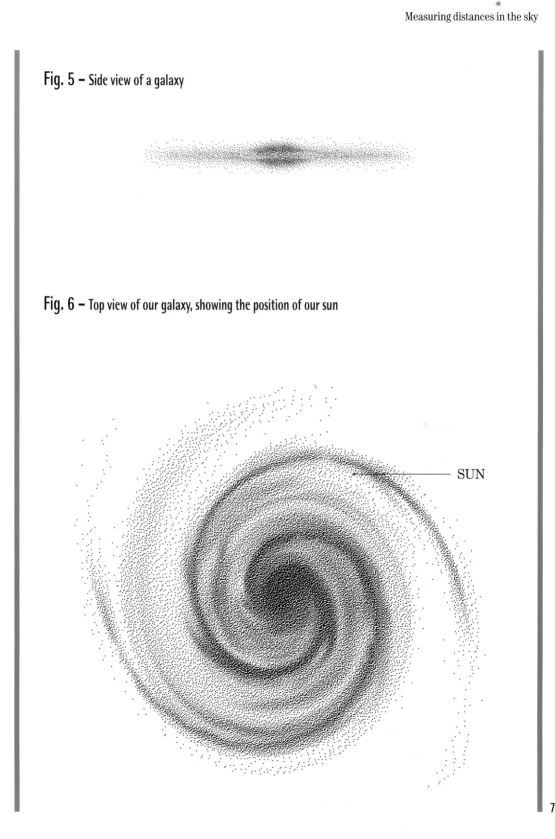

Fig. 6 – Top view of our galaxy, showing the position of our sun

SUN

Imagine yourself sitting near the end of a spiral arm of our galaxy. If you look straight up or down you will see neighboring stars in our spiral arm of our galaxy but when you look toward Sagittarius you are looking along the flat side of the spiral arm toward the wider and more dense center bulge of our galaxy. We cannot see the spiral arm opposite us because it is hidden by the billions of stars in the center of the galaxy.

As you look along the Milky Way you will notice that some areas appear to have dark holes or slots in them. These are not empty spaces, but rather dark masses of dust, star debris and gases that hide the stars behind them. There is a very definite dark slit in the area of Cygnus, sometimes called the Cygnus Rift, or the Northern Coalsack, and a similar dark patch in area of the Southern Cross, called the Southern Coalsack.

Life in the heavens

There are nine known planets orbiting around our Sun. There are over 200 000 000 000 (200 billion) suns (stars) in our galaxy. There are billions of galaxies. Just imagine how many planets there must be in our galaxy alone. It is therefore almost impossible to assume that we are the only planet with life on it. In 1995, three planets have almost certainly been identified in distant stars. Two of them appear to be as far from their Sun as Earth is from our Sun. This suggests that at least as far as distance to the life-giving source of a Sun is concerned they are not unlike Earth. It is also important to realize

that the basic elements necessary for life as we know it, carbon, hydrogen, oxygen and nitrogen, exist throughout the heavens. The question is not whether there are any living organisms among the stars, but rather what kind of life is there and are they trying to contact us? We should realize that amino acids have been found in meteorites. Given the proper environmental conditions these molecules may join to form the proteins and RNA of living cells, which can then replicate themselves. Such action signifies life.

Instructions for use of the atlas

Begin your walk through the sky by first determining which constellations are visible overhead during the month of your walk. Therefore, look at the following four star charts, which give an overview of the constellations visible during each of the four seasons. Look at the chart for the appropriate season. Hold the chart in front of you like reading a book. Do not place it overhead. You will notice north at the bottom. Face north and compare the lower half of the chart with the stars in the sky. Then face south, east, or west and turn the star chart around so that the direction you are facing is at the bottom of the chart.

When you have determined which constellation you wish to see, turn to the index to find out which are the diagrams for that constellation. There is a date at the top of each diagram. If you are looking at the sky on a different date, the positions of the constellations will not change in relationship to each other. Those relationships are constant, but since the stars appear to move due to the Earth's rotation, you may have to turn the diagram in order for the constellations to appear as they do at your time and date.

If you are not familiar with the constellations then start your walk with the Big Dipper on page 14. Concentrate upon the four main constellations, Ursa Major, Ursa Minor, Cepheus and Cassiopeia (see Fig. 5 in Part 2). If you are familiar with some of the star groups, then, depending upon the season, look for a specific constellation by using the index to find the diagram dealing with that constellation's relationship to other constellations. If you are only slightly familiar with some constellations the four charts here can help you find others.

Spring Stars

March-April 10-12 pm
May-June 7-9 pm

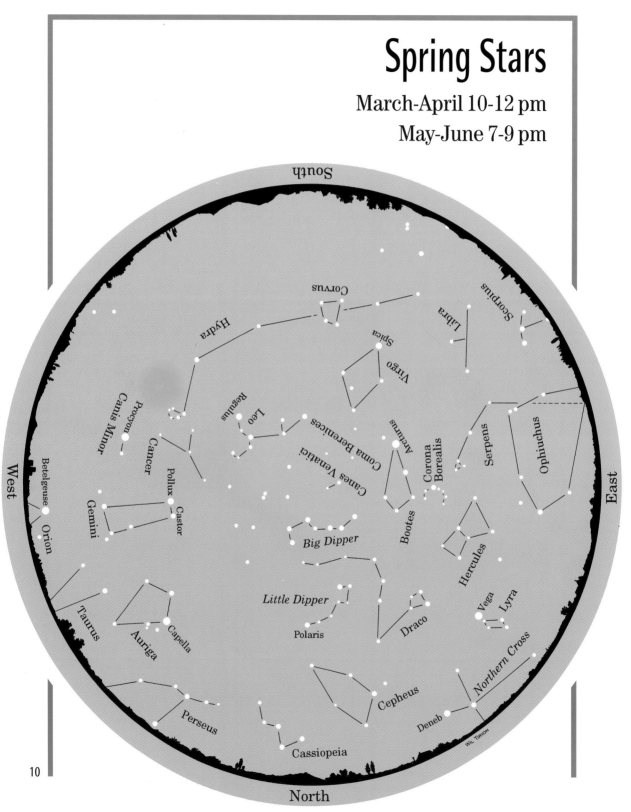

Summer Stars

June–July 10–12 pm
August–September 7–9 pm

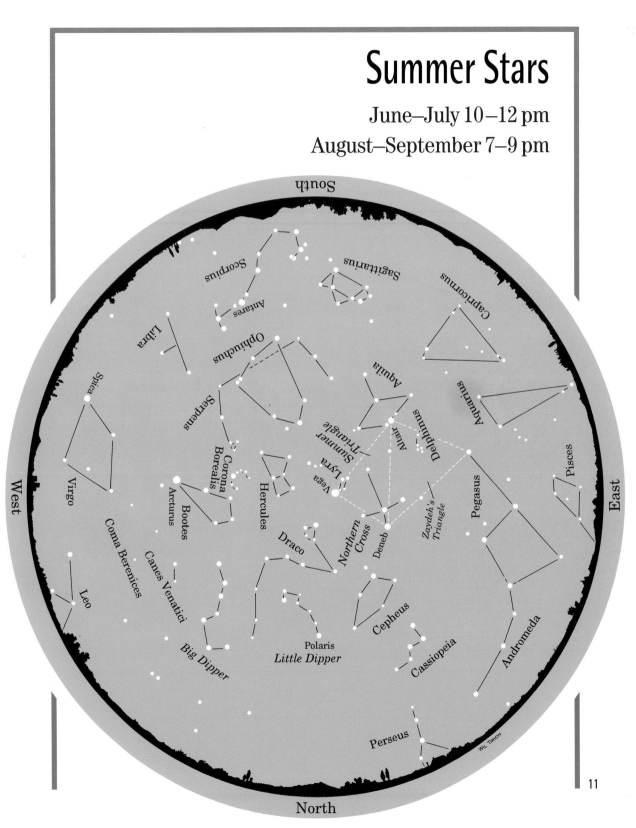

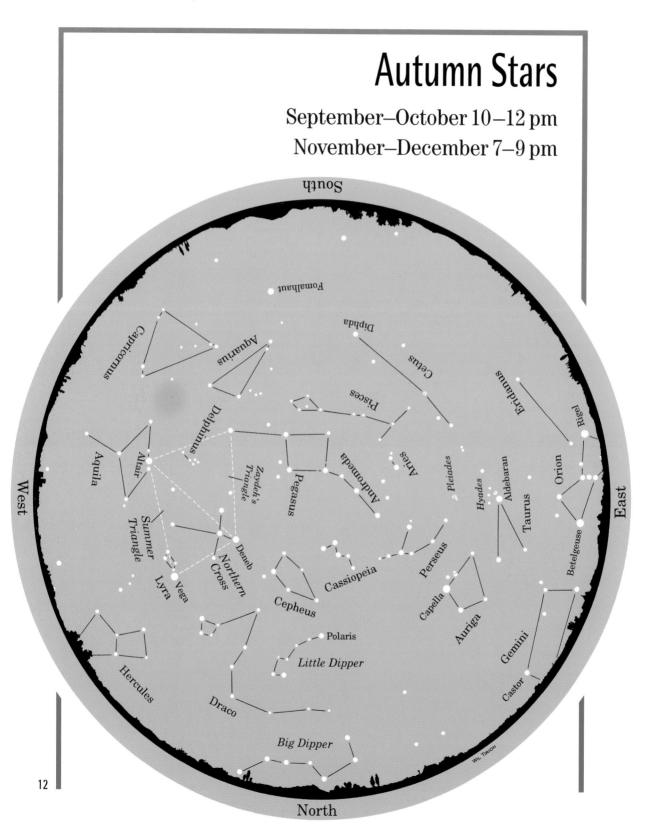

Autumn Stars

September–October 10–12 pm
November–December 7–9 pm

Winter Stars

December–January 10–12 pm
February–March 7–9 pm

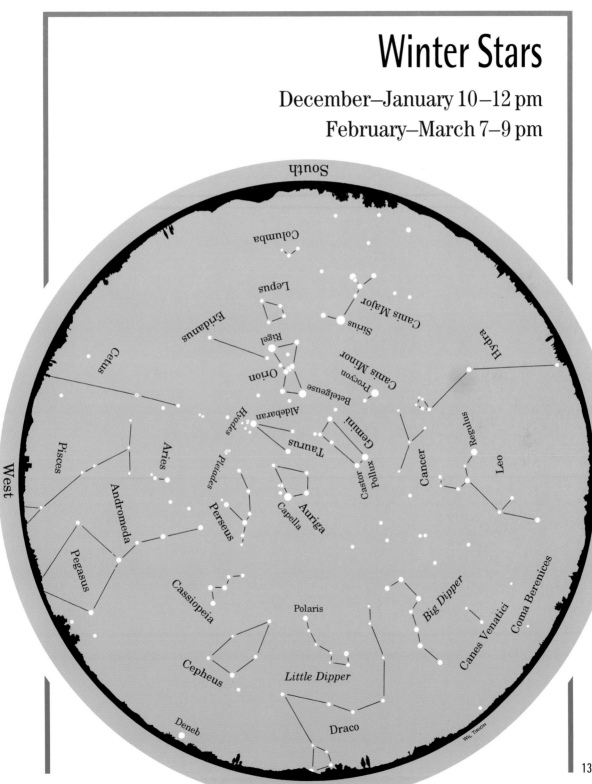

South

Columba

Lepus

Eridanus

Canis Major

Sirius

Rigel

Hydra

Cetus

Orion

Betelgeuse

Canis Minor

Procyon

Regulus

Hyades

Aldebaran

Pisces

Aries

Taurus

Gemini

Cancer

Leo

West

Pleiades

Castor

Pollux

East

Andromeda

Perseus

Capella

Auriga

Pegasus

Cassiopeia

Big Dipper

Coma Berenices

Polaris

Canes Venatici

Cepheus

Little Dipper

Deneb

Draco

WIL TIRION

North

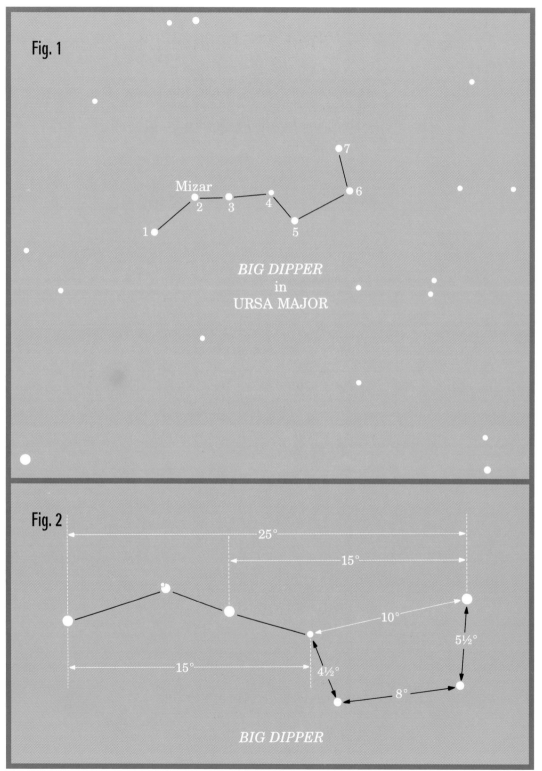

Fig. 1

Mizar

2

3

4

5

6

7

1

BIG DIPPER
in
URSA MAJOR

Fig. 2

25°

15°

15°

10°

5½°

4½°

8°

BIG DIPPER

Part 2   A walk through the heavens

Let us start our walk by locating the Big Dipper, which is part of the larger constellation Ursa Major. The Big Dipper is best seen between January and October, when it is not too close to the horizon. It is also known as the Plough and as the Cooking Pot.

Look toward the north and look for seven bright stars shaped like a large ladle or saucepan as in Fig. 1. It may be on its side or upside down, depending on the season. There is one star whose name you should remember, Star 2 is Mizar. Stars 6 and 7 are called the pointers since they point to Polaris, the North Star. Once you have found the Big Dipper look at it very closely. Ignore the other stars. Measure its size with your hand.

How close do you come to the measurements in Fig. 2?

For convenience, each star in each constellation will be numbered so that we can more easily identify specific stars to help us walk around the sky. We will follow several paths to a constellation. By doing this you will have a better sense of star positions and relationships.

Let us begin our walk with the Big Dipper. By using only the Big Dipper, depending upon the season, we will be able to locate many different constellations or specific stars within those constellations.

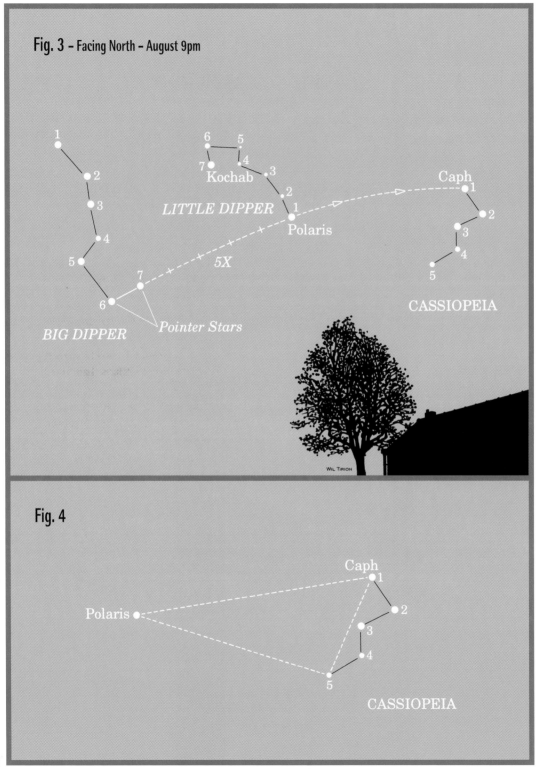

Fig. 3 – Facing North – August 9pm

LITTLE DIPPER

Kochab

Polaris

5X

BIG DIPPER

Pointer Stars

Caph

CASSIOPEIA

WIL TIRION

Fig. 4

Polaris

Caph

CASSIOPEIA

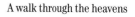

Starting from the Big Dipper

To locate Polaris in the Little Dipper (Fig. 3)

☆ Follow Stars 6 and 7 of the Big Dipper to Polaris, the North Star. It is five times the distance between 6 and 7.

The Little Dipper has only two bright stars, Polaris (Star 1), and Kochab (Star 7). Polaris is about 0.8 degrees away from the celestial north pole, which is the point in the sky to which the north– south axis of the Earth is pointing. The Norse people believed that there was a huge spike through Polaris around which the universe revolves.

To locate Cassiopeia

Cassiopeia consists of five bright stars. Notice that the W or M shape spreads out at one end. Stars 3 and 5 are farther apart than Stars 1 and 3. This is important to recognize.

☆ A line from Stars 6 and 7 of the Big Dipper through Polaris continues in a slight curve to the big W or M in the sky.

☆ A straight line from Star 1, 2, or 3 of the Big Dipper, through Polaris, leads to Cassiopeia.

☆ To find Polaris from Cassiopeia form a triangle joining Star 1 and Star 5 of Cassiopeia toward Polaris (see Fig. 4). Star 5 is closer to Polaris than Star 1.

It is good to learn the names of some of the stars as we go along. They become much friendlier, so from now on I will refer to Star 1 of Cassiopeia by its proper name Caph.

Fig. 5 – Facing North – August 9pm

To locate Cepheus (Fig. 5)

The stars of Cepheus are not very bright, but they are not hard to find. Stars 1, 2 and 3 are the brightest. Cepheus resembles a house with a tall roof.

Imagine a slight curve between Polaris and Caph of Cassiopeia. About one-third of the way between them, closer to Polaris, find a faint star. This is Star 1, the peak of the house of Cepheus. The house image is upside down in the diagram.

To find Star 3 of Cepheus, imagine a line between Polaris and Star 3 of Cassiopeia. This becomes the base of a triangle. Now form a triangle with almost equal sides by moving past Caph to a mildly bright star which becomes the third point of the triangle. Its name is Alderamin.

Between Alderamin and Polaris is Star 2 of Cepheus. We now have the peak and one side of Cepheus. Now complete the house.

At this time stop and review the four constellations, the Big Dipper, the Little Dipper, Cassiopeia and Cepheus. Do not look for other constellations until these are clear in your mind's eye. Remember the three star names, Polaris of the Little Dipper, Caph of Cassiopeia and Alderamin of Cepheus.

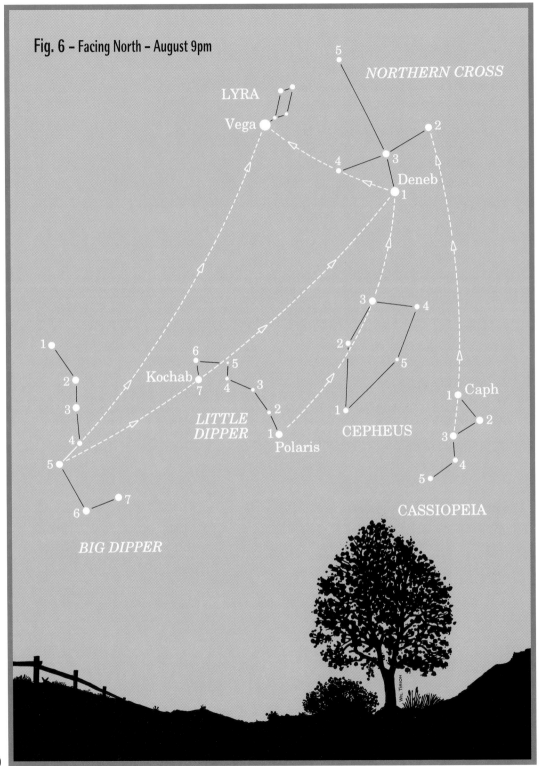

Fig. 6 – Facing North – August 9pm

NORTHERN CROSS

LYRA

Vega

Deneb

LITTLE
DIPPER

Kochab

Polaris

CEPHEUS

Caph

CASSIOPEIA

BIG DIPPER

To locate the Northern Cross (Cygnus) (Fig. 6)

Cygnus, the Swan, lies along the Milky Way. Star 1 (Deneb), a supergiant, is 60000 times brighter than the Sun, but is 1500 light years away.

- Go from Star 3 of Cassiopeia through Caph (Star 1) to Star 2 of Cygnus, which is the tip of the crossbar of Cygnus.

- A line from Star 5 of the Big Dipper through Star 7 (Kochab) of the Little Dipper goes directly to Deneb (Star 1 of Cygnus).

- A line from Polaris through Alderamin (Star 3) of Cepheus goes to Deneb of the Northern Cross. Now complete the cross.

To locate Vega in Lyra

Vega is one of the brightest stars in the sky (magnitude 0.3).

- Go from Deneb (Star 1) of Cygnus in a slight curve past Star 4 of Cygnus to the very bright star Vega.

- A line from Star 2 of Cassiopeia through Caph (Star 1) goes in a slight curve to Vega.

- Star 5 of the Big Dipper through Star 4 of the Big Dipper is almost a straight line to Vega.

Study the Northern Cross and Vega relationship and review their relationship to the Big Dipper before you go any further.

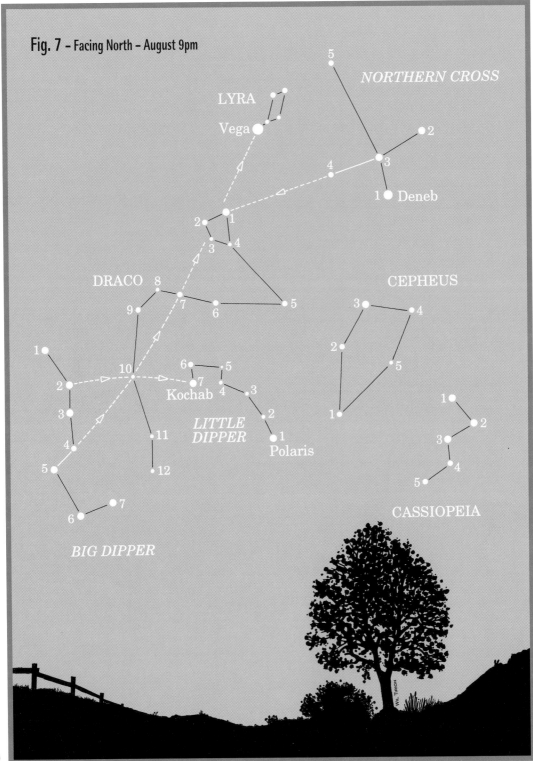

Fig. 7 – Facing North – August 9pm

NORTHERN CROSS

LYRA

Vega

Deneb

DRACO

CEPHEUS

LITTLE DIPPER

Kochab

Polaris

BIG DIPPER

CASSIOPEIA

WIL TIRION

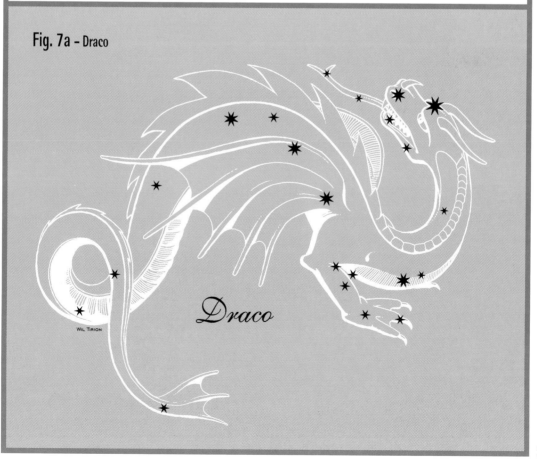

To locate Draco (Fig. 7)

Draco, the dragon, is a long winding constellation.

☆ Star 3 of the crossbar of the Northern Cross through Star 4 points directly to Star 1 of Draco's head, which is an irregular square. It is three times the distance between Stars 3 and 4 of the Northern Cross.

☆ A line from Star 5 of the Big Dipper through Star 4 goes directly through the head of Draco and continues in a straight line to Vega of Lyra.

☆ Thuban (Star 10) in Draco's body is faint and lies along a soft curve between Mizar (Star 2) of the Big Dipper and Kochab (Star 7) of the Little Dipper.

Several thousand years ago Thuban was the pole star. The bright star (Star 1) in the head of Draco was called Isis, after an Egyptian goddess. The doors of the holy chambers of the temples of Thebes and Denerach, which were built to honor her, were aligned to Isis.

Fig. 7a - Draco

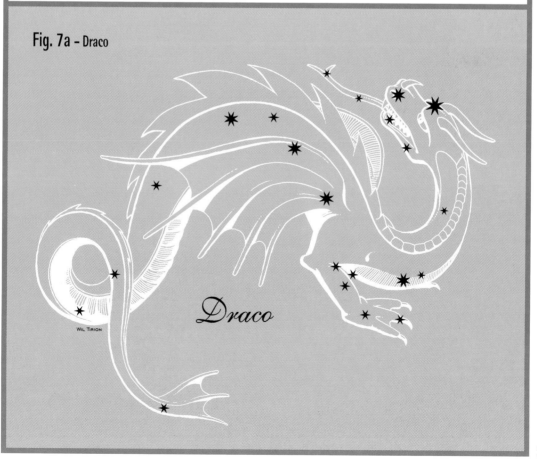

Draco

Wil Tirion

Fig. 8 – Facing South – May 9pm

BIG DIPPER

CORONA BOREALIS

CANES VENATICI

Cor Caroli

Alphecca

BOOTES

Arcturus

Wil Tirion

To locate Arcturus in Bootes (Fig. 8)

Bootes looks like a giant kite. In 1933 the light from Arcturus, the fourth brightest star in the sky, which is approximately 36 light years away, was focused onto a photo-electric cell which produced electrical current. This current was amplified and then used to open the gate of the 1933 World's Fair in Chicago, Illinois. That light had left Arcturus in 1897!

When Arcturus is seen early in the evening it signifies the coming of spring.

 Follow the curve of the handle of the Big Dipper to the bright star Arcturus. It is approximately 30 degrees away from the end of the handle. If you continue the arc past Arcturus it will lead you to Spica in Virgo (see Fig. 14).

To locate Cor Caroli in Canes Venatici

A line from Star 7 of the Big Dipper going between Stars 4 and 5, but much closer to Star 5, leads to Cor Caroli, which is approximately 15 degrees from Star 5 of the Big Dipper.

To locate Corona Borealis

Corona Borealis, the Northern Crown, is a charming constellation. It lies between Bootes and Hercules, but is closer to Bootes (see Fig. 9). Alphecca, Star 3, is called The Jewel in the Crown. The six other stars are very faint.

Move from Arcturus to Star 3 of Bootes. Alphecca is just to the side of this star.

Fig. 9 – Facing South– June 9pm

1 ● Deneb

2 · 3 · 4

5

NORTHERN
CROSS

LYRA

Vega

Head of
DRACO

HERCULES

4 · 5
3 · 6
2
1

BOOTES

4 · 5
3
2
1
Arcturus

7
1
2
6
5 4 3

CORONA
BOREALIS

Rasalhague ●
in OPHIUCHUS

WIL TIRION

To locate Hercules (Fig. 9)

Hercules is a faint constellation which lies between Lyra and Corona Borealis.

☄ A line from Star 4 of Cygnus through Vega goes to Star 2 of Hercules. This distance is about 11/2 times the distance between the tip of the crossbar of Cygnus and Vega.

☄ A line from Star 3 of Cygnus through Vega goes to Star 3 of Hercules.

☄ From Arcturus through Alphecca, the bright star of Corona Borealis (Star 3), to Star 6 of Hercules is a straight line. Now construct the bent hourglass shape of the six stars of Hercules.

Most of Hercules lies within a giant triangle formed by Vega of Lyra, Arcturus of Bootes and Rasalhague of Ophiuchus (see Fig. 24).

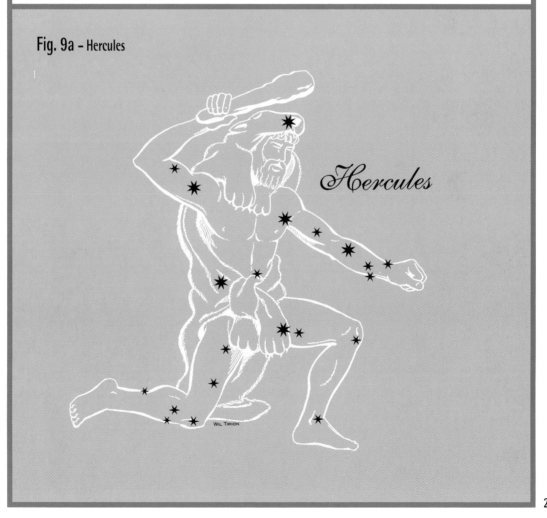

Fig. 9a - Hercules

Hercules

WIL TIRION

Fig. 10 – Facing South – November 9pm

CASSIOPEIA

AURIGA

Capella

Mirfak

Almak

ANDROMEDA

Algol

PERSEUS

Pleiades

TRIANGULUM

PEGASUS

Hamal

ARIES

Aldebaran
in TAURUS

PISCES

WIL TIRION

To locate Perseus and Andromeda (Fig. 10)

Perseus crosses the Milky Way.

⭐ Move from Star 3 of Cassiopeia through Star 4 in a slight curve approximately five times the distance between Stars 3 and 4 to the bright star Mirfak (Star 2) in Perseus.

⭐ Go from Caph (Star 1) of Cassiopeia through Star 2 directly to Almak (Star 1) of Andromeda. The distance is approximately four times the distance between Caph and Star 2.

⭐ Algol (Star 5 of Perseus) almost forms a right-angled triangle with Mirfak and Almak.

Mirfak and Algol are the two brightest stars in Perseus.

Now visualize the curve of five or six stars of Perseus which leads to the Pleiades going south, or to Cassiopeia going north.

Andromeda lies between Perseus and Pegasus. It contains the only major galaxy outside of our own galaxy that we can see with our naked eye (see Figs. 3 and 7a of Part 4).

To locate Pegasus (Figs. 10 and 11)

There are several ways to find the giant square of Pegasus, the winged horse.

⭐ Follow a line in a soft curve from Mirfak in Perseus to Almak of Andromeda past Star 2 to Star 3 of Andromeda, which is the same as Star 1 of Pegasus. This is one corner of the square of Pegasus.

⭐ A line from Star 3 of Cassiopeia to Star 2 points directly to the center of Pegasus. This distance is approximately 35 degrees.

⭐ See next page and Fig. 11.

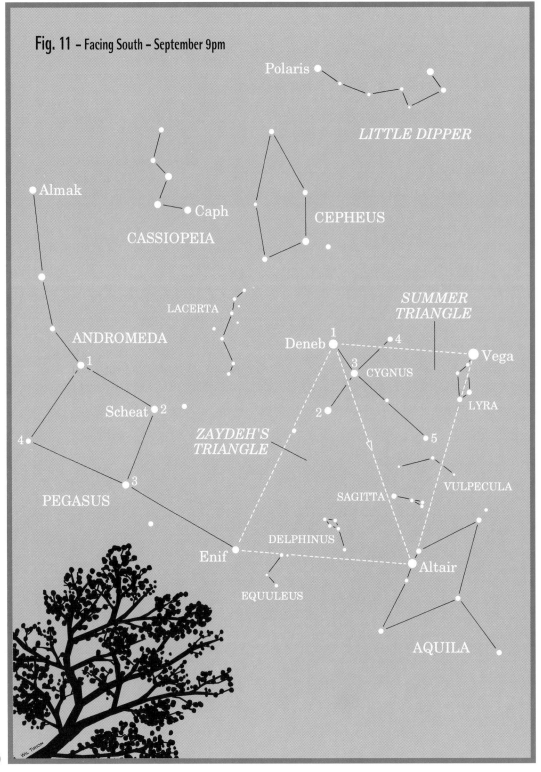

Fig. 11 – Facing South – September 9pm

Polaris

LITTLE DIPPER

Almak

Caph

CASSIOPEIA

CEPHEUS

LACERTA

SUMMER
TRIANGLE

ANDROMEDA

Deneb

Vega

1

3 4
CYGNUS

2

LYRA

Scheat 2

1

ZAYDEH'S
TRIANGLE

5

4

VULPECULA

3

SAGITTA

PEGASUS

DELPHINUS

Enif

Altair

EQUULEUS

AQUILA

WIL TIRION

To locate Pegasus (Fig. 11)

To find Pegasus from the opposite direction, we should be aware of the Summer Triangle. This important star relationship is formed by Deneb of Cygnus (the Northern Cross), Vega of Lyra, and a third bright star, Altair of Aquila.

To locate Aquila

☆ A line from Deneb going between Stars 2 and 3 of Cygnus leads to Altair in Aquila. Aquila is a diamond-shaped constellation with a tail at the corner of the diamond opposite Altair. Altair is its only bright star.

Look closely at Altair, notice the faint stars on either side of it.

Now study the *Summer Triangle*, which contains the constellations of Sagitta and Vulpecula.

To locate Pegasus

☆ A line from Star 3 of Cygnus between Deneb and Star 2 goes directly to Star 1 of Pegasus.

☆ A line through the crossbar of Cygnus points in a slight curve to Enif in Pegasus.

☆ Join Deneb, Altair and Enif to form Zaydeh's triangle opposite the Summer Triangle. This triangle is an easy way to identify Enif and then to move from Altair through Enif in an upward curve to find the square of Pegasus.

Zaydeh's triangle contains Delphinus and meets Equuleus.

Study all these relationships to get a good image of the giant square of Pegasus, which is over 15 degrees wide.

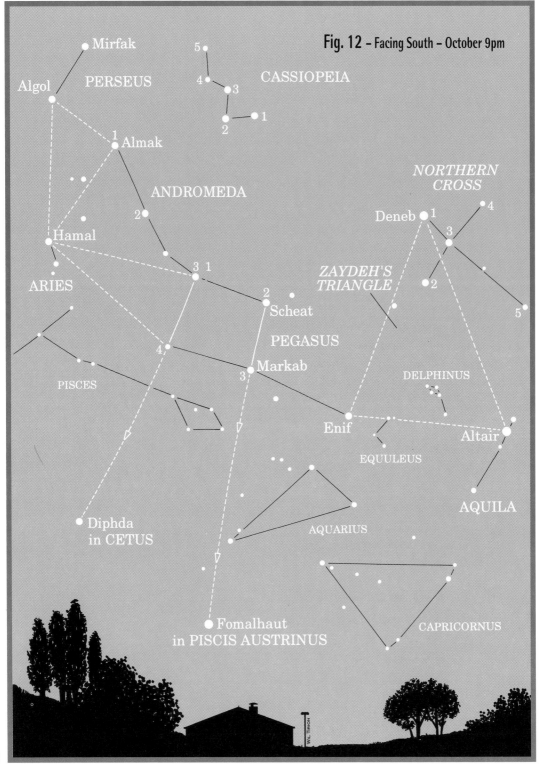

Fig. 12 – Facing South – October 9pm

Mirfak

PERSEUS

Algol

CASSIOPEIA

5
4
3
2
1

1 Almak

ANDROMEDA

NORTHERN CROSS

Deneb
1
4
3
2
5

2

Hamal

3 1

ZAYDEH'S TRIANGLE

ARIES

2

Scheat

PEGASUS

PISCES

DELPHINUS

4

3 Markab

EQUULEUS

Enif

Altair

AQUILA

Diphda in CETUS

AQUARIUS

Fomalhaut in PISCIS AUSTRINUS

CAPRICORNUS

WIL TIRION

32

To locate Hamal in Aries, Diphda in Cetus and Fomalhaut in Piscis Austrinus (Fig. 12)

To locate Aries
From Perseus:

☆ Algol of Perseus and Almak of Andromeda form a triangle with the star Hamal of Aries.

From the square of Pegasus:

☆ Star 1 and Star 4 of Pegasus, when joined to Hamal of Aries, form a large isosceles triangle.

From Orion:

☆ Find Aldebaran in Taurus then move through the Pleiades and continue in a soft curve downward approximately 25 degrees to the moderately bright star Hamal of Aries (see Figs. 10 and 23).

To locate Fomalhaut in Piscis Austrinus
Fomalhaut is one of the brightest isolated stars in the southern sky.

☆ A line from Star 2 (Scheat) of Pegasus through Star 3 (Markab) leads to Fomalhaut, which is approximately 50 degrees from Star 3.

To locate Diphda in Cetus

☆ A line from Star 1 of Pegasus through Star 4 leads to Diphda, which is to the side of Fomalhaut.

Study the relationship between Pegasus, Pisces, Aquarius, Capricornus, Delphinus and Aquila.

Fig. 13 – Facing South – May 9pm

BIG DIPPER

To Arcturus

COMA BERENICES

LEO

2
1
3
4
5
9
Regulus
6
7
8
Denebola

Arcturus

Alphard

SEXTANS

VIRGO

HYDRA

Spica

CRATER

LIBRA

Kraz

CORVUS

1
2
3 SCORPIUS
Antares

Wil Tirion

To locate Spica in Virgo (Fig. 13)

Follow the arc of the handle of the Big Dipper past Arcturus and continue the arc to a bright star, which is Spica. Find the irregular diamond shape of Virgo.

A line from Antares of Scorpius through Star 2 of Scorpius goes directly through Libra in a curve to Spica.

Spica is part of an equilateral triangle with Arcturus and Denebola of Leo.

To locate Corvus

Follow the curve of the handle of the Big Dipper to Arcturus continuing past Spica to locate Kraz, which is Star 1 in Corvus.

To locate Hydra, the Water Snake

This serpent, which was slain by Hercules, is the longest constellation in the sky, 100 degrees long. During springtime it extends across the lower southern sky below Virgo, Corvus, Leo and Cancer. It actually cuts across Crater and Corvus. There is only one bright star – Alphard.

A line from Star 4 of Leo through Star 6 (Regulus) leads directly to Alphard.

A line from Castor through Pollux of Gemini leads past the head of Hydra (see Fig. 15) to Alphard (see Fig. 13).

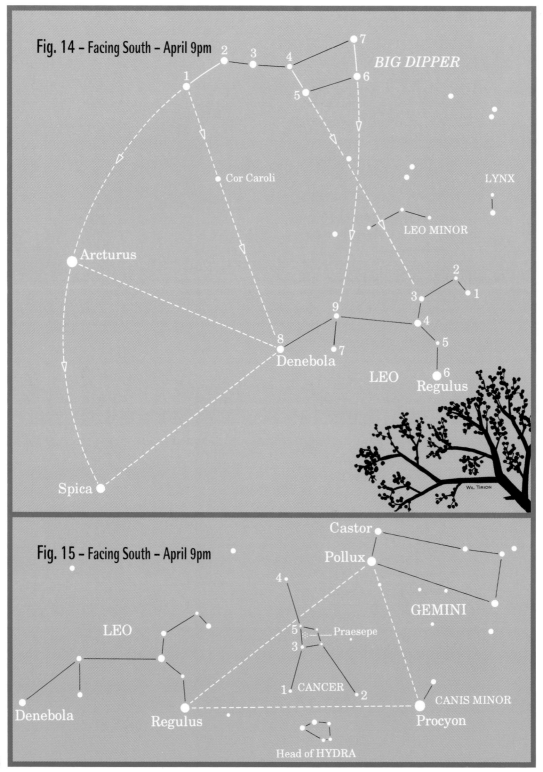

Fig. 14 – Facing South – April 9pm

BIG DIPPER

LYNX

LEO MINOR

Cor Caroli

Arcturus

LEO

Denebola

Regulus

Spica

WIL TIRION

Fig. 15 – Facing South – April 9pm

Castor

Pollux

GEMINI

LEO

Praesepe

CANCER

Denebola

Regulus

Head of HYDRA

CANIS MINOR

Procyon

To locate Leo (Fig. 14)

Leo looks like a toy horse. The head, chest and front legs look like a backwards question mark. There are two stars in Leo whose names you should learn. Star 6 is Regulus, which means Little King. Star 8 is Denebola, the tip of Leo's tail.

✦ A line from Star 4 of the Big Dipper through Star 5 leads to Star 3 of Leo, and in a slight curve to Regulus. Star 3 of Leo is approximately 40 degrees from Star 5 of the Big Dipper.

✦ A line from Star 7 of the Big Dipper through Star 6 leads to Star 9 of Leo, the beginning of the tail.

✦ Form a triangle joining Arcturus of Bootes, Spica of Virgo and Denebola of Leo.

✦ A line from Betelgeuse of Orion to Procyon of Canis Minor (see Fig. 21) leads in a slight curve to Regulus (see Fig. 15).

✦ Form a triangle joining Pollux of Gemini, Procyon of Canis Minor and Regulus (Fig. 15).

To locate Cancer (Fig. 15)

The very faint constellation of Cancer lies in the center of the triangle formed by Procyon of Canis Minor, Pollux of Gemini and Regulus of Leo. In the center of the square of Cancer is a faint cluster of stars called the Beehive or Praesepe.

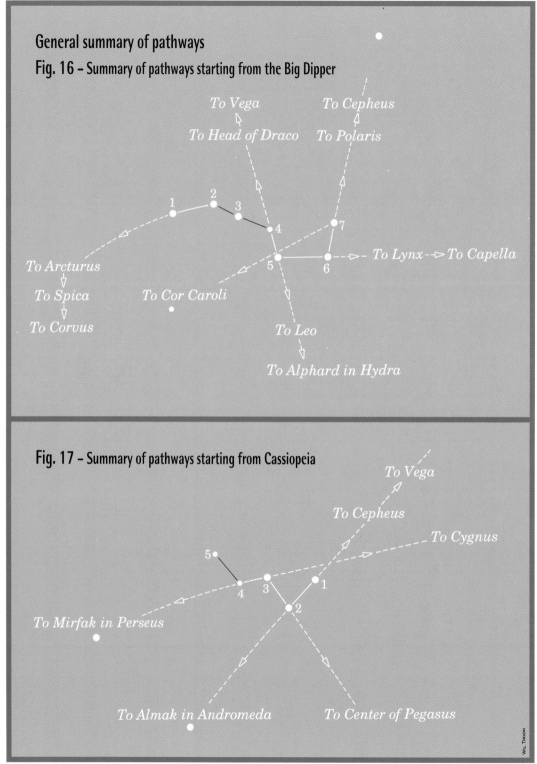

General summary of pathways

Fig. 16 – Summary of pathways starting from the Big Dipper

To Vega

To Head of Draco

To Cepheus

To Polaris

To Arcturus

To Spica

To Corvus

To Cor Caroli

To Lynx --▷ To Capella

To Leo

To Alphard in Hydra

Fig. 17 – Summary of pathways starting from Cassiopeia

To Vega

To Cepheus

To Cygnus

To Mirfak in Perseus

To Almak in Andromeda

To Center of Pegasus

WIL TIRION

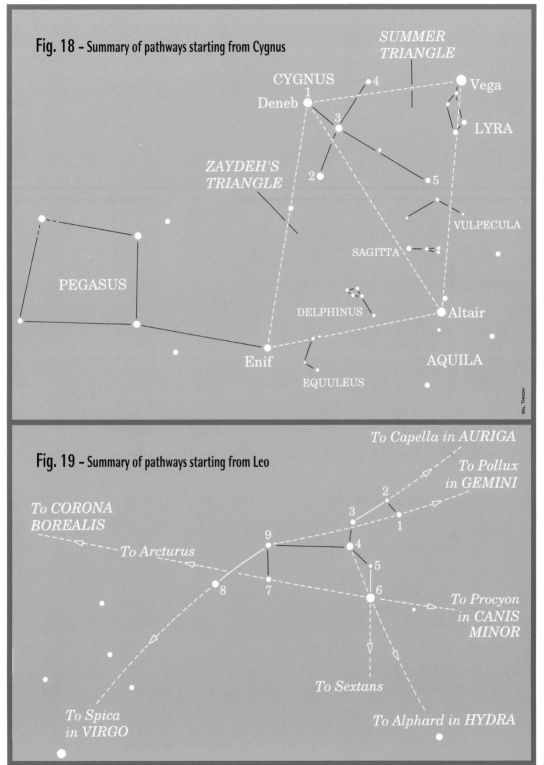

Fig. 18 – Summary of pathways starting from Cygnus

SUMMER TRIANGLE

CYGNUS

Deneb

Vega

LYRA

ZAYDEH'S TRIANGLE

VULPECULA

SAGITTA

PEGASUS

DELPHINUS

Altair

Enif

AQUILA

EQUULEUS

Wil Tirion

Fig. 19 – Summary of pathways starting from Leo

To Capella in AURIGA

To Pollux in GEMINI

To CORONA BOREALIS

To Arcturus

To Procyon in CANIS MINOR

To Sextans

To Spica in VIRGO

To Alphard in HYDRA

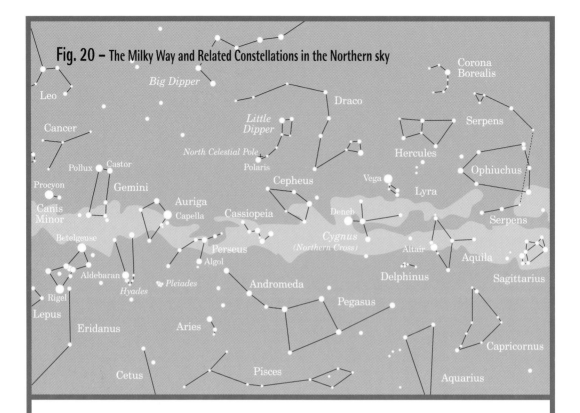

Fig. 20 – The Milky Way and Related Constellations in the Northern sky

The Orion area

Half of Orion is surrounded by six very bright stars (see Fig. 21).

Sirius of Canis Major
Procyon of Canis Minor
Pollux of Gemini
Castor of Gemini
Capella of Auriga
Aldebaran of Taurus

Do not be confused if you see an unusually bright 'star' in this region. It may not be a star. It may be a planet.

Starting from Orion

Orion, the Hunter, may be seen in the southern sky during the winter months, from late September to mid April. Orion is a large striking constellation. Look for seven bright stars in the shape of an hourglass with three bright stars (Stars 3, 4, 5) in a row forming the waist of the hourglass or the belt of the hunter. Star 1 is Betelgeuse and Star 7 is Rigel. See Fig. 21a.

Once you have identified Orion, test your ability to measure distances. Stars 6 and 7 are approximately 7 degrees apart. The distance between Stars 1 and 7 is 20 degrees. Study Orion closely since it serves as an excellent guide to the surrounding constellations and to the Pleiades.

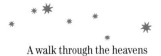

A walk through the heavens

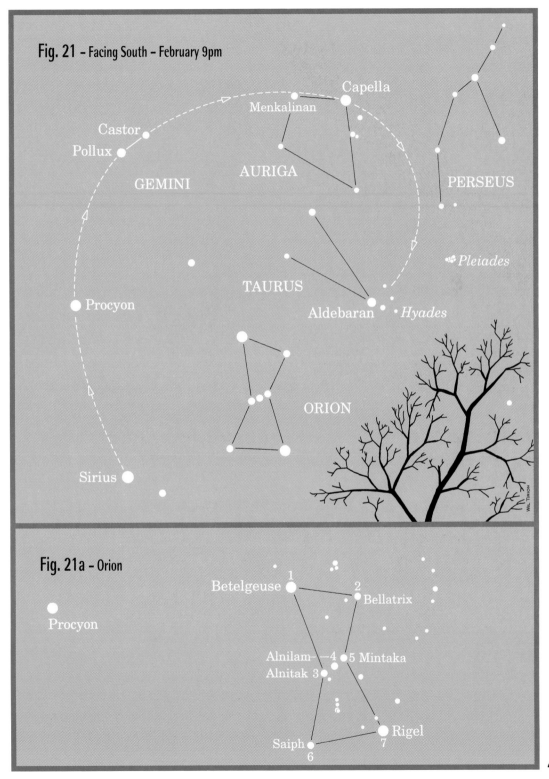

Fig. 21 – Facing South – February 9pm

Capella

Menkalinan

Castor

Pollux

GEMINI

AURIGA

PERSEUS

Pleiades

TAURUS

Procyon

Aldebaran *Hyades*

ORION

Sirius

WIL TIRION

Fig. 21a - Orion

Betelgeuse 1   2 Bellatrix

Procyon

Alnilam——4 5 Mintaka
Alnitak 3

Saiph 6   7 Rigel

41

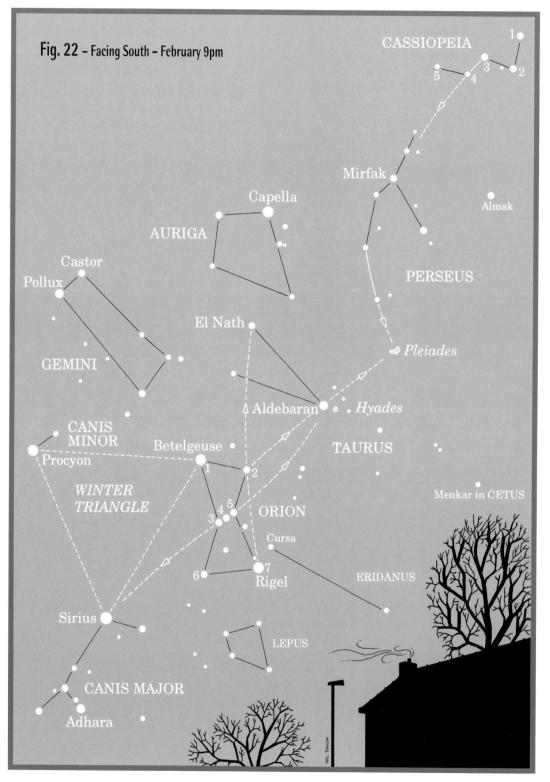

To locate Sirius in Canis Major (Fig. 22)

Sirius is the brightest star in the sky.

 A line through the belt of Orion (Stars 3, 4, 5), passes just above Sirius on one side and just below Aldebaran of Taurus on the other side. Sirius is about 15 degrees from Star 6 (Saiph) of Orion. Aldebaran is 15 degrees from Star 2 (Bellatrix) of Orion.

To locate Procyon in Canis Minor

This is a bright star 25 degrees higher in the sky from Sirius. You may notice that the Milky Way lies between Procyon and Sirius.

The Winter Triangle is an equilateral triangle joining Star 1 (Betelgeuse) of Orion, Sirius of Canis Major and Procyon of Canis Minor.

To locate Taurus and its star clusters the Hyades and the Pleiades

Taurus is the symbol of springtime, the time for planting and the symbol of love.

A line from Rigel (Star 7) of Orion through Bellatrix (Star 2) of Orion leads directly to El Nath of Taurus and then almost in a straight line to Capella in Auriga.

Stars 3, 4 and 5 of Orion point in a slight curve to Aldebaran.

To locate the Pleiades

It is easier to find the Pleiades from Orion in the south, but it is a good exercise to locate it by walking across the northern sky from Cassiopeia through Perseus. You will then become more familiar with the relationship between the northern and southern constellations.

From Orion walk past Aldebaran approximately 10 degrees to a faint hazy cluster of stars, the Pleiades.

From Cassiopeia move to Mirfak of Perseus and then along the curve of Perseus to the Pleiades.

It now becomes easy to find Polaris, the North Star, from Orion. Walk from Orion's belt to Aldebaran, to the Pleiades, to the arc of Perseus, to Cassiopeia, to Polaris (see Fig. 4).

Read about the Pleiades in the legends of Taurus and Ursa Major.

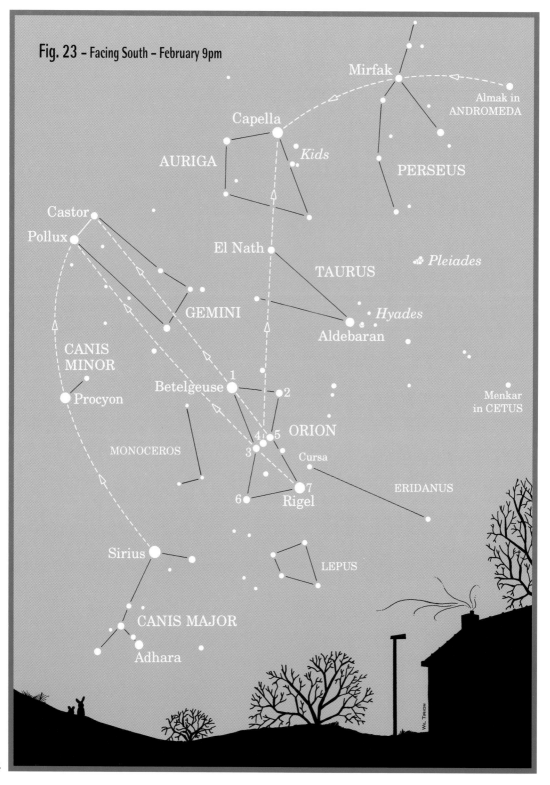

Fig. 23 – Facing South – February 9pm

Mirfak

Almak in
ANDROMEDA

Capella

Kids

PERSEUS

AURIGA

Castor

Pollux

El Nath

TAURUS

Pleiades

GEMINI

Hyades

Aldebaran

CANIS
MINOR

Betelgeuse

Procyon

1

2

Menkar
in CETUS

ORION

4 5

3

Cursa

MONOCEROS

ERIDANUS

6

Rigel

7

Sirius

LEPUS

CANIS MAJOR

Adhara

WIL TIRION

To locate Gemini (Fig. 23)

Gemini forms a nice rectangle that includes the stars Pollux and Castor, which are frequently referred to as the Twins. They symbolize true friendship.

- A curve from Sirius to Procyon leads to Pollux and Castor.

- A line from Star 7 (Rigel) of Orion through Star 3 leads to Pollux.

- Walk from Star 5 of Orion through Star 1 (Betelgeuse) in a slight curve downward to Castor, which is approximately 30 degrees from Star 1.

To locate Auriga

Auriga, the Charioteer, is shaped like a kite. It lies above the horns of Taurus. Capella is the sixth brightest star in the sky. Just below and to the right of Capella are three faint stars, called the Kids (baby goats).

- An arc from Almak of Andromeda through Mirfak of Perseus leads to the bright star Capella.

- A line from Star 5 through Star 6 of the Big Dipper leads directly to Capella, a distance of approximately 50 degrees.

- A line from Vega through Polaris passes near Capella.

- A walk from Star 4 of Orion going between Star 2 and Star 1 passes through El Nath of Taurus to Capella.

Capella is 23 degrees from Aldebaran and from Castor. It is 10 degrees from Mirfak and 35 degrees from Betelgeuse. Test your ability to measure these distances.

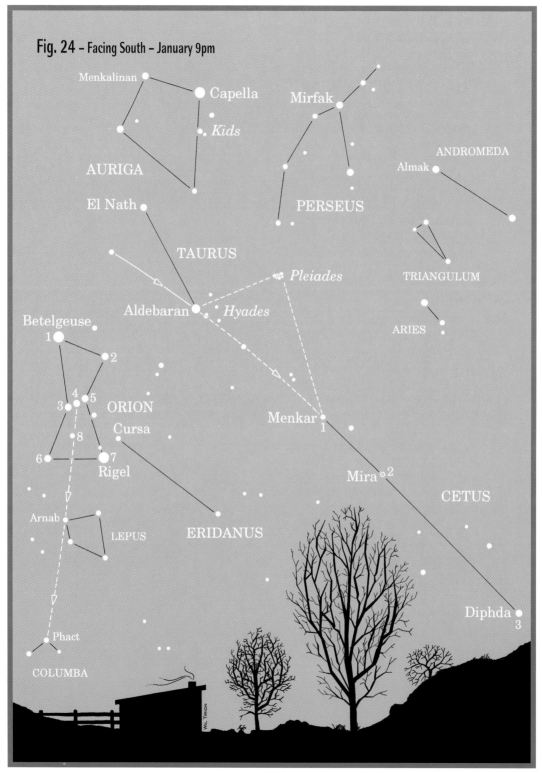

Fig. 24 – Facing South – January 9pm

To locate Lepus, Columba, Eridanus and Cetus (Fig. 24)

A line from Star 4 of Orion through Star 8 leads to Arnab of Lepus. This line continues to Phact of Columba.

Stars 6 and 7 of Orion form a triangle with Arnab in Lepus.

To locate Eridanus

Approximately 3 degrees above and to the right of Rigel (Star 7) of Orion is the star Cursa of Eridanus.

To locate Cetus

This constellation has three fairly bright stars, Menkar, Star 1, Mira, Star 2 (which is a variable star – its brightness fluctuates) and Diphda, Star 3.

A large isosceles triangle joins the Pleiades and Aldebaran of Taurus and Menkar of Cetus.

The tip of the bottom horn of Taurus (Star 3) through Aldebaran (Star 1) of Taurus curves down slightly to Cetus – 25 degrees from Aldebaran.

Now return to Orion (Fig. 21) and locate Sirius again. Study the circle of stars going from Sirius – to Procyon – to Pollux – to Castor – to Menkalinan in Auriga – to Capella and then curving downward to Aldebaran.

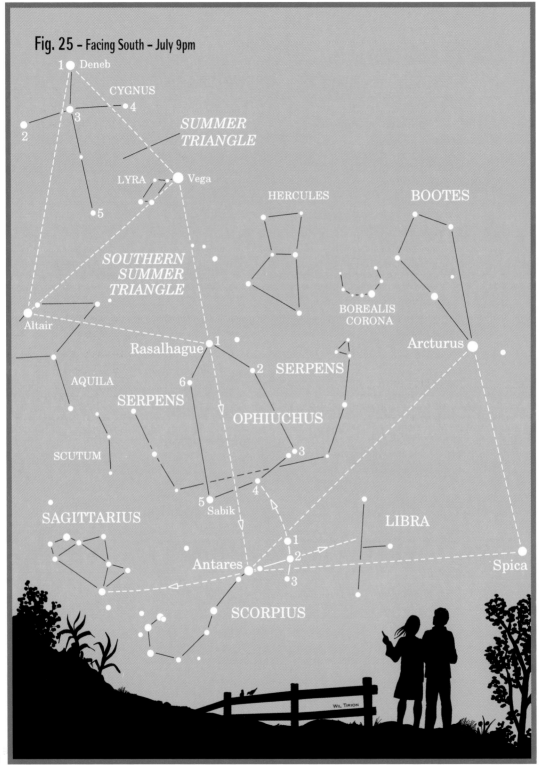

Fig. 25 – Facing South – July 9pm

Deneb

CYGNUS

SUMMER TRIANGLE

LYRA

Vega

HERCULES

BOOTES

SOUTHERN SUMMER TRIANGLE

BOREALIS CORONA

Altair

Rasalhague

SERPENS

Arcturus

AQUILA

SERPENS

OPHIUCHUS

SCUTUM

Sabik

SAGITTARIUS

LIBRA

Antares

Spica

SCORPIUS

WIL TIRION

To locate Ophiuchus (Fig. 25)

Ophiuchus is known as the Serpent Bearer.

- On the base of the Summer Triangle form an equilateral triangle by joining Altair, Vega and Rasalhague (Star 1) of Ophiuchus. This is the 'Southern Summer' Triangle

- The arc of Scorpius 3–2–1 leads to Star 4 of Ophiuchus.

Rasalhague (Star 1) is about 30 degrees from Sabik (Star 5).

To locate Serpens

Serpens is a divided constellation which appears to pass through the lower part of Ophiuchus. It lies above Scorpius.

To locate Antares in Scorpius

Scorpius, which looks like a large scorpion, is bright and low in the sky. Its brightest star Antares is flanked by a star on either side.

- A line from Vega of Lyra to Rasalhague of Ophiuchus goes in a mild curve to Antares.

- A large triangle is formed by joining Arcturus of Bootes, Spica of Virgo and Antares.

To locate Sagittarius

Sagittarius, the Teapot, lies between our solar system and the center of our galaxy. The Milky Way is most dense behind Sagittarius.

- A line from the mid star of the head of Scorpius (Star 2) through Antares leads to the bottom star of Sagittarius.

To locate Libra

- A line from Antares of Scorpius through Star 2 in the head of Scorpius leads directly to Libra, the Scale of Justice.

Now study the relationship between Altair of Aquila, Scorpius, Sagittarius, Cygnus, Vega of Lyra, Arcturus of Bootes and Spica of Virgo.

Part 3 Legends of the Heavens

Legend of Andromeda

See the legends of Cassiopeia and Perseus.

Legend of Aquarius

The legend of Aquarius is about the water carrier Ganymede. Ganymede was a shepherd boy who was so kind and gentle that he was given ambrosia, the food of the gods, to make him immortal. One day, while tending his sheep and playing with his dog Argos, the god Jupiter/Zeus sent Aquila, his giant eagle, to sweep down to the plains of Troy to take Ganymede to the temple of the gods to become Jupiter's favorite water carrier. Wherever Jupiter went Ganymede would accompany him by riding on the back of Aquila the eagle. Ganymede's kindness was again made evident to the gods when he asked Jupiter if he could help the Earth people, who were in need of water. Jupiter, who was usually not very kind, was softened by Ganymede's compassion, and gave him permission to do as he wished. Ganymede realized that to send a great deal of water to Earth at one time may be dangerous so he decided to send it in the form of rain. That is how Ganymede, the shepherd boy, became known as the god of rain.

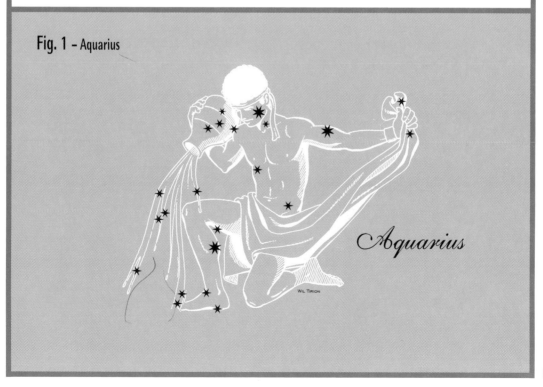

Fig. 1 – Aquarius

Aquarius

WIL TIRION

Legend of Aquila

Aquila was Zeus's pet Eagle. Aquila was not only involved with Ganymede in the legend of Aquarius and how Earth was given rain, but was also part of the story of how people got fire.

The Titans were giant gods who were fighting the Greek god of Olympia, Zeus, the new ruler of the world. Prometheus, one of the Titan gods, did not oppose Zeus during the war. After the Titans were beaten he became an advisor to Zeus. While serving Zeus he became aware that the Earth people did not have fire and were not only suffering from the cold but could not have warm food. He felt sorry for them and therefore stole a ray of sunshine, hid it in a bamboo container, and sent it down to Earth. With this ray of sun the Earth people made fire to warm their bodies.

Zeus became very angry when he saw that the Earth people were given fire without his permission. He captured Prometheus and chained him to the side of a mountain in the Caucasus where he was to remain forever, and to suffer repeated attacks by Aquila the eagle. After every attack the wound would heal and then would be ripped open again by Aquila. One day when Aquila was about to bite into the abdomen of this kind chained Titan god, Hercules, who agreed with Prometheus's act of kindness, and who was angered at what Jupiter had done, shot one of his magic arrows into Aquila. Aquila fell seriously wounded. Zeus healed the wounds of his pet eagle and then placed it in the heavens so that it may still soar. It flies near the tail of Cygnus the swan.

Legend of Aries

Aries was the pet ram of Zeus, the Greek ruler of the heavens. Its coat was made of golden fleece instead of white wool. One day Zeus was looking down on the Earth people when he suddenly noticed that two children on Earth were in danger of being killed. He immediately sent Aries down to Earth to save them. Aries arrived just in time for the children to jump on his back as he raced to safety. To honor the effort his ram had made, Zeus placed him in the heavens where he can roam freely near the flying horse Pegasus.

Aries also symbolizes the ram caught in the thicket where Abraham was about to sacrifice Isaac. In gratitude to God who sent the angel to stop the hand of Abraham, the ram was placed upon the altar and sacrificed to honor God.

Legend of Auriga

Auriga is portrayed as the guardian of the shepherds, who carries a goat in his arms as he rides through the sky on his chariot. Shepherds all over the world know that whenever the constellation of Auriga appears in the sky the rains will soon fall, the grasslands will grow, and the sheep will have all the food they need.

It is said that the god Jupiter accidentally broke a horn of one of the goats. He apologized for this accident by filling the broken horn with good things. Such a horn has been called the 'Horn of Plenty' or a cornucopia.

Legend of Bootes

Bootes was the son of Demeter, the Greek goddess of agriculture. He was a sensitive young man with a great sense of purpose and social consciousness. When he saw the Earth people struggling to find food he wanted to help them. He realized that if he were to send them food they would always need his help. Instead, he decided to help them to be able to help themselves, to be independent, so he built the first plough and sent it to Earth. Since then people have been able to plow the land and grow their own food. Because of this great deed for mankind the gods honored him by placing him in the heavens near the Big Dipper (which is also known as the Plough).

Legend of Canis Major and Minor

Canis Major and Canis Minor were the hunting dogs of Orion. Canis Major was so swift a runner that it could overtake any animal. It was therefore greatly valued by Orion.

The early Egyptians saw the bright star Sirius in Canis Major as the god Anubis, the god with a man's body and the head of a jackal. When Sirius appeared in the sky before dawn, it was the time of the flooding of the Nile, which was of great importance to the farmers who lived along the river since the flood always brought new silt to the land and replenished the soil. It became known as the Dog Star, and the hot days of summer, between July and early September, became known as the Dog Days.

Legend of Cassiopeia

Cassiopeia, the wife of King Cepheus, was the beautiful queen of Ethiopia. She was so proud of her beauty that she became arrogant. She even boasted that her beauty was greater than that of the sea nymphs, the Nereids. This boast angered the sea nymphs, who were the daughters of the sea god Nereus, not because they were so vain, but because Cassiopeia failed to appreciate that her external beauty was something she was born with and not something she had achieved. Gratitude for her good fortune would have been acceptable, but not pride. To be proud of something which was not gained through personal effort, but rather with which you were born, showed a poor sense of values.

The Nereids asked the ruling god of the sea, Poseidon (Neptune), to punish Cassiopeia because of her distorted sense of values and her conceit. Poseidon therefore ordered the giant sea monster, Cetus, to destroy their kingdom of Ethiopia.

When King Cepheus and Queen Cassiopeia were informed of Neptune's decision they went to the old wise oracle of Ethiopia for advice. He told them that they must sacrifice their lovely daughter Andromeda to appease the sea gods. Although they were heartbroken, they chained Andromeda to a rock on a cliff overlooking the sea knowing that Cetus, the sea monster, would destroy her.

When Cetus, the sea monster, began to race toward her she screamed for help. Meanwhile, Perseus, who was on the great winged horse Pegasus returning home with the head of Medusa (see the legend of Perseus), heard her cry and immediately flew to her rescue. He arrived just in time to hold up the head of Medusa as Cetus approached. The sea monster was immediately stopped, since anyone who looked directly at the head of Medusa was turned to stone. Perseus carefully placed the head back in its sack, taking care that Andromeda did not look at it. He then unchained

Andromeda who fell into his arms. When they gazed into each other's eyes they immediately fell in love.

Although the sea god was angry because the punishment he decreed was not fulfilled, he was so touched by the love of Perseus and Andromeda that he placed them next to each other in the heavens so that their love will always be seen and felt by us on Earth.

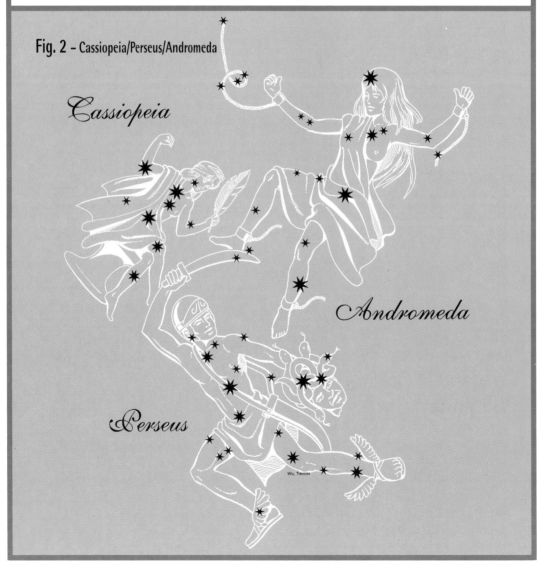

Fig. 2 – Cassiopeia/Perseus/Andromeda

Cassiopeia

Andromeda

Perseus

WIL TIRION

Poseidon felt that Cassiopeia deserved more punishment. He therefore placed her in the sky in a position where she is condemned to circle around the celestial pole forever, half of the time in the upside-down position.

Legend of Cetus

Read the legends of Perseus and Cassiopeia.

Legend of Coma Berenices

Berenice was a beautiful woman whose hair was most glorious. She was married to the Egyptian King Evergetes. When the king left on a dangerous mission Bernice vowed to dedicate her hair to the goddess of beauty if the king were to return unharmed. When he finally returned safely she cut her beautiful hair which Jupiter then placed among the stars. It appears as a cluster of faint stars with a lacy appearance that lies near Arcturus of Bootes and Cor Coroli of Canes Venatici. Coma Berenices has become a symbol of the sacrifices that everyone should be willing to make for their loved ones.

Legend of Corona Borealis

Among the North American Indians this constellation is considered to be a council of chiefs sitting in a semicircle to discuss the future of their people.

In ancient Greece the story is told of Ariadne the beautiful daughter of King Minos of Crete. When her lover, who was mortal, found out that she had been promised to be wed to a god, he left her on the island of Naxos. Bacchus, the god of vegetation and wine, saw her, fell in love and asked her to marry him. Ariadne did not believe Bacchus was a god. To prove to her that he was a god, Bacchus asked Venus, the Goddess of love, to design a crown of magnificent jewels as a wedding present for Ariadne. When Venus did this for Bacchus, Ariadne was convinced that he was a god and consented to marry him. Bacchus was so ecstatic with joy that he threw the crown of jewels into the heavens where it has been shining ever since.

Legend of Corvus

Corvus was the pet crow of Apollo, the god of sun and music. He was a magnificent bird with a beautiful song. One day Apollo sent Corvus on an errand with instructions to return without delay. Corvus immediately left on his mission. On his return home he saw a fig tree with unripe fruit. This tantalized him, so he waited by the tree for several days until the figs were ripe enough to eat. After eating them he hurried back to Apollo. When asked why it took so long to return, Corvus made up an excuse, but Apollo knew the excuse was false. He was very disappointed that Corvus, whom he had trusted, was not honorable enough to tell him the truth. As punishment Apollo changed the beautiful song of the crow to a hoarse caw sound.

Next time you see a crow think of the constellation of Corvus and the reason for Apollo's punishment.

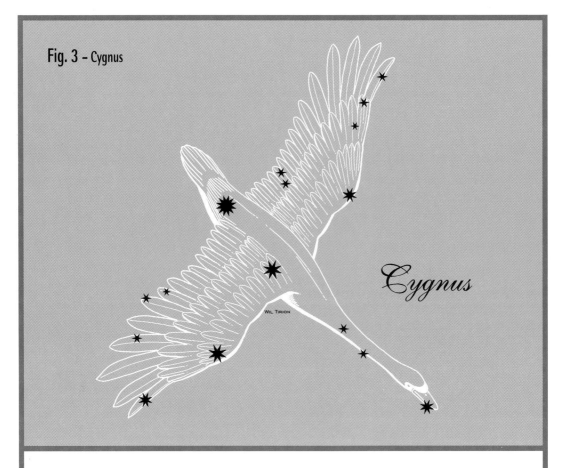

Fig. 3 - Cygnus

Cygnus

WIL TIRION

Legend of the Crater

The Crater (the Cup) is next to the constellation of Corvus the crow. The Egyptians were very aware of the Crater. They knew that when the Crater rose above the horizon the river Nile, which had been flooding the Egyptian plain, would not rise further and would soon begin to recede.

Legend of Cygnus

Cygnus the swan symbolizes the wonder, the goodness and the dedication which exists in true friendship. Both the constellations of Cygnus and Gemini symbolize the significance of friendship. Cygnus and Phaethon were very close friends. Phaethon, the son of the mortal woman Clymene, pleaded with his father Helius, the sun god, to help him convince the Earth people that he was the son of a god. Helius agreed to help and therefore told Phaethon that he would grant him any wish. Phaethon immediately asked for permission to drive the four winged horses pulling the chariot carrying the sun. His father pleaded with him not to ask for the almost impossible task of controlling the winged horses, but Phaethon insisted that his father keep his promise. As dawn neared he mounted the chariot with great excitement and began to race across the sky. The great winged horses sensed the inexperienced control of the reins and raced so fast that Phaethon lost complete control of the horses. The chariot swayed so much that the sun was about to fall out of the chariot and burn the Earth. The god Zeus saw what was happening and in order to save the Earth from being destroyed by the heat of the sun, threw a thunderbolt at the chariot. Phaethon lost his balance and fell off the chariot into the roaring river Eridanus. Cygnus saw his friend disappear into the river, and immediately, in spite of the danger, dove into the waters to save him. Helius was so overwhelmed by this act of true friendship toward his son that he changed Cygnus into a diving swan flying along the line of the Milky Way as a symbol of the greatness and importance of friendship.

Legend of Delphinus the Dolphin

Approximately 2600 years ago, on the island of Lesbos lived a man by the name of Arion. He was a famous poet who was also endowed with a magnificent voice. Arion performed in concerts throughout Greece and Italy. On one of his trips he was sailing to his home in Corinth Greece with a group of valuable prizes. When the sailors realized the value of the prizes they became very greedy and decided to steal the wealth and throw Arion overboard. When Arion realized what they were going to do he pleaded with them to allow him to sing one more song while playing his lyre. They

agreed. He sang a beautiful song of gratitude to Apollo the god of music and poetry for blessing him with such wonderful talents. Apollo heard the song and knew what was going to happen. He immediately asked the sea god Poseidon to send his messengers, the dolphins, to surround the boat. As Arion sang he noticed the unusual number of dolphins suddenly swarming around the boat. He then jumped overboard with his lyre before the sailors had a chance to grab him. As he was sinking into the deep sea the largest dolphin dived under him, raised him to the surface and then surrounded by the other dolphins they raced away carrying Arion to safety. When Apollo heard about the magnificent action of the dolphin he wished to honor it. He therefore placed it in the heavens and placed the lyre of Arion nearby in the constellation of Lyra.

Legend of Draco the Dragon

The ancient Chaldeans, who lived in the region of the Euphrates and Tigris rivers, believed that the dragon Tiamat, the monster of chaos, darkness and evil, lived before the sea and the sky were separated. Tiamat was challenged by the light of the sun and the gods who arose out of the sea of chaos. But he was so powerful and frightening that even the gods gave way. Evil appeared to be winning until Marduk, one of the gods of light, appeared. He had been given all the magical powers that the gods of light and goodness could bring together, and with this power he overcame the dragon, light gained over darkness and good over evil. Tiamat the dragon was placed in the sky as Draco to show all gods and all people that goodness can win.

Legend of Eridanus

The river of Eridanus flows from its origin near Rigel of Orion to flow under Taurus toward Cetus the monster whale. See the legend of Cygnus the swan.

Legends of Gemini

Among the Maori of New Zealand is the tale of twin brothers who were the mortal children of Bora bora. The brothers were extremely devoted to each other and preferred to play with each other rather than with other children. This relative isolation disturbed their parents who then decided to separate the boys. The twins overheard their parent's discussion and decided to run away. They sailed away but their mother followed them. They went from island to island but she was always behind them until they reached Tahiti where they hid in the mountains. She discovered their hiding place and was about to capture them when they climbed to the top of a mountain and flew to the sky where they will always remain close together.

In ancient Greece there is the legend of Castor, a famous horse tamer and soldier, and Pollux, a champion boxer, were the sons of the Greek god Zeus. They were not only brothers, but very close friends and very adventuresome. At one time they decided to go to sea in order to attack pirates who had been raiding honest seamen. They were so successful in their war against pirates that they became warlike heroes to the sea people, who honored them by carving their images on the prow of their ships. Seamen are aware that during stormy weather sparkling lights may appear on the rigging. When two sparks appear it is an omen that Castor and Pollux are protecting the ship and that the ship will weather the storm. These lights have been called St. Elmo's Fire.

During one of their fights with thieves Castor, who was mortal, was killed. Pollux, who was immortal, was grief stricken and begged Zeus to allow him to be with Castor every other day in the underworld. Zeus was so touched by this request and by Pollux's feeling of true friendship that he not only approved of the request, but placed them in the heavens together so that the Earth people would always be reminded of the preciousness of true friendship.

Legend of Hercules

Hercules, the son of the god Zeus/Jupiter and of the beautiful mortal woman Alcymene, was the greatest of ancient Greek heroes. He began to show his great physical strength as a young child, but more importantly, Hercules revealed his fine sense of character as a young man when he met two women called Pleasure and Virtue. Pleasure promised him enjoyment while Virtue promised him hard work, but glory as a doer of great deeds to help mankind. He chose Virtue and was subsequently taught by the wise centaur Chiron.

Fig. 4 - Gemini

Gemini

WIL TIRION

His deeds included ridding the world of monsters. He fought for 30 days with Leo the Nemean lion before he was able to kill it. He then destroyed the enormous nine-headed water snake of Lernea, which would capture and eat those who ventured near its swamp. The snake was then thrown into the sky and is represented by the constellation Hydra. While fighting the water snake he also destroyed the giant crab, which is now in the sky as Cancer. He captured the wild boar which was destroying the vineyards as well as the fire-breathing bull which was devastating the land.

Hercules continued his work until he had performed 12 deeds. Several years later he was poisoned by mistake with the blood of a centaur. When he died the gods raised him into heaven where he can still be seen as a symbol of one who is dedicated to helping all people by doing good deeds.

Legend of Hydra the Water Snake

Hydra was a nine-headed snake-like monster that killed and ate people as they traveled near the swamps of Lernea. Its blood was poisonous. Hercules was asked to destroy the monster. During their fight every time Hercules would cut off a head another would immediately grow back in its

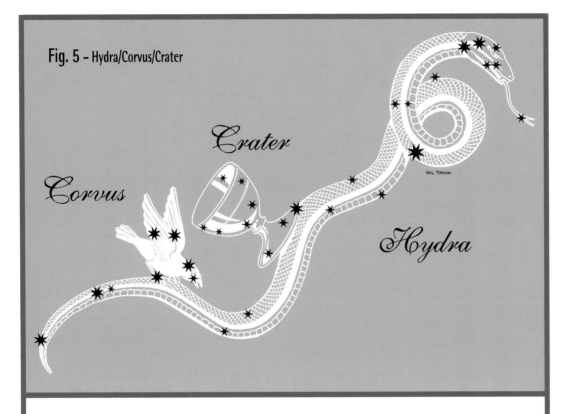

Fig. 5 - Hydra/Corvus/Crater

Crater

Corvus

Hydra

Wil Tirion

place. He finally succeeded in killing the monster by burning the cut surface on the neck, which prevented regrowth of a head.

Legend of Leo

Leo was a lion who lived on the moon. Food was scarce so he tried to attack one of the horses pulling the chariot of the moon goddess Selene. Leo was thrown off and landed on Earth near Nemea in Greece where he began to attack people. Hercules was called to destroy this destructive lion. He made a huge knotty club and approached the lion's den. When the lion attacked he swung and struck the lion on the tip of its nose. The lion retreated into its cave, but fearless Hercules was

determined to destroy it, so he followed it into the cave. The roof of the cave was so low that Hercules could not use his club, so he jumped on the lion's back and strangled it with his bare hands.

This heroic act was seen by Zeus who honored Hercules by placing the conquered lion in the sky. See the legend of Hercules.

Legend of Libra

Libra symbolizes Astraea, the goddess of justice. She would weigh the souls of men and women on a balancing scale and hold them responsible for their acts. The Sumerians in 200 BCE called it the 'Balance of Heaven'.

Legends of Lyra

Among the Maori the brightest star in Lyra was called Whanui and symbolizes a legend of a love triangle. One night Whanui met the beautiful wife of Rango-Maui. Her name was Pani. Whanui was so overcome by her beauty that although he knew he would be doing wrong he seduced Pani and made love to her. She subsequently gave birth to the sweet potatoes. Her husband Rango-Maui was so disturbed by their presence that Pani allowed him to send the sweet potato children down to Earth. This so angered Whanui that he, in retaliation, sent three kinds of caterpillars down to earth to feed upon the sweet potatoes. As a result, before Whanui, the brightest star in Lyra, appears in the sky at dawn the Earth people store sweet potatoes in the ground to avoid the caterpillars.

Lyra represents the lyre, a harp, which the Greek god Hermes invented and made from a tortoise shell. Its sound was glorious but Hermes was unable to make it sing, so he gave it to Apollo, his brother. Although Apollo was able to make it sing he could not make it sound soulful regardless how much he tried, so he called Orpheus, who was a great musician, to test the harp. When Orpheus picked up the instrument and moved his fingers across the strings the Earth seemed to become silent. All things were listening, the beasts, the birds, the trees and even the flowers turned their faces toward Orpheus. When Apollo saw how music affected all living things he gave the lyre to Orpheus who would play so that people would feel the uplift of music when life seems difficult.

There are times, when the night is very dark and still, that one may look up at Lyra and possibly hear the murmur of Orpheus's song among the sounds of the night.

Legend of Ophiuchus

Among the Babylonians the stars of Ophiuchus and Serpens were thought to portray the sun god Marduk fighting with the dragon Tiamat (Draco). But later in Greek mythology, Ophiuchus was identified with Aesculapius, the Greek god of medicine.

Aesculapius was the son of Apollo and the Thessalian princess Coronis, who died giving birth to Aesculapius. When he was a youth he had such a radiant appearance that everyone knew that he had to be one of the gods. Chiron, the wisest of the Centaurs, taught him the art of medicine. One day Aesculapius observed a snake carrying a herb in its mouth which was used to revive another snake that had been killed. Aesculapius took the herb and with it expanded his knowledge of medicine.

Aesculapius was becoming so knowledgeable that the god Zeus feared he would learn how to defeat the death of mortals. To prevent this Zeus felt that Aesculapius must die. He regrettably destroyed him with a bolt of lightning, but then placed him among the stars in the constellation of Ophiuchus. Since then Aesculapius and the snake have been special symbols of healing. Hippocrates, upon whose name all physicians swear an oath to respect the sick, was supposedly a descendant of Aesculapius.

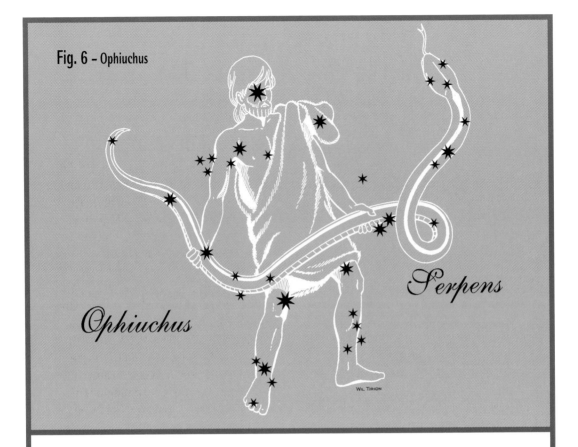

Fig. 6 – Ophiuchus

Ophiuchus

Serpens

Wil Tirion

Today's emblem of medicine, the Caduceus, commemorates this legend. It is a winged staff with two snakes entwined around it, similar to the staff which was carried by the Greek god Hermes.

Legend of Orion

In the northern tip of Australia in Arnhem Land the Aborigines of Yolngu speak of the time when three famous fishermen, who belonged to the kingfish totem, spent several days at sea trying to catch fish. They were successful, but only in catching kingfish, which they could not eat since it was the totem of their people. They were in a terrible dilemma for their children would go hungry if they did not return with some fish. In desperation they decided to break the taboo against eating kingfish. They resumed fishing and soon caught three more kingfish. The Sun, amazed and angered that they would kill and eat their totem, called upon the clouds, the sea and the wind to create a gigantic waterspout. It was so powerful that it whirled the three fishermen and their canoe high into the sky. To this day they may be seen seated in their canoe as the three stars in a row in Orion. If one looks very carefully, just below the three stars you may be able to see the tiny fish hanging below their canoe.

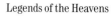

Fig. 7 - Orion

Orion

WIL TIRION

one wish. Hyrieus, who was childless, asked to have a son. The wish was granted and Orion was born. Orion grew up and became a superb hunter for he had been blessed by the gods, but as he became more and more famous as a great hunter he also became insensitive to the animals he hunted. He actually enjoyed the killing of an animal. He did not hunt and kill for necessity. He was so unfeeling about the life of animals that Artemis, the goddess of hunting, sent the giant scorpion (Scorpius) to attack him. He was stung and about to die when Ophiuchus the healer gave Orion an antidote which saved his life. When Orion recovered he realized, after being so close to death, how precious life is and how pitiless and uncaring he had been. He repented and therefore was placed in the heavens with the scorpion whose sting had taught him that all life is precious. See the legend of Sagittarius.

Among the Ju/Wasi people of Africa is the legend of the god Old/Gao who was hunting for zebras. He finally saw three of them lined up in a row; took aim and shot his arrow, but missed his target. The three zebras escaped and now may be seen as the three center stars of Orion. The arrow may still be seen where it fell. It lies just below the three zebras facing away from them.

In Greek mythology, Hyrieus, a poor farmer who lived in Thebes, was a kind man who frequently befriended strangers although he was poor. One day he helped three unusual strangers. He did not know that they were the gods Zeus, Neptune and Hermes. In return for his kindness he was granted

Legend of Pegasus

Pegasus, the magical winged horse of ancient Greek mythology, was the offspring of Poseidon and Medusa. He helped Perseus race through the sky to save beautiful Andromeda (see the legend of Cassiopeia). He was also the steed who rode on the wind carrying the hero Bellerophon through his adventures.

Bellerophon was the son of Corinth and the grandson of Sisyphus. Sisyphus was a selfish and arrogant man who took advantage of people who were less clever. He was therefore punished by the gods and forced to roll a massive stone to the top of a hill, but whenever the top was almost reached the stone would always slip and fall to the bottom. He was forced to continue this struggle for the rest of his life.

Bellerophon was wrongly accused of doing something evil. He was therefore sent on several dangerous missions which he was able to accomplish with the help of Pegasus. As a result he was given the right to keep Pegasus.

As Bellerophon grew older he became arrogant like his grandfather and too proud of his possession of a magical horse that could even ride to the gods. Although Bellerophon was only mortal, he tried to force Pegasus to take him to the top of Mount Olympus so he could mingle with the gods. Pegasus was so astounded at this arrogance that he reared up and threw Bellerophon off his back. Bellerophon fell to Earth while Pegasus flew to the gods.

Whenever we see the constellation of Pegasus it should remind us that gentleness and the doing of good deeds is usually rewarded, while arrogance and selfishness may lead to failure and destruction.

Legend of Perseus

King Acrisius of Argos in Greece was told by an oracle that he would someday be killed by his grandson. To prevent this he imprisoned his daughter Danae so that no one could reach her. But Jupiter saw her and fell in love with her. The prison was no barrier. When his daughter gave birth to Perseus, the king put them both in a chest and set them adrift on the sea.

The chest did not sink, but eventually landed safely on the island of Seriphus controlled by King Polydectes. When Perseus became a young man he was full of adventure and eager for glory. King Polydectes fell in love with Danae and realized that her devotion to Perseus would interfere with his courtship of her. He therefore asked Perseus to bring him the head of the Gorgon Medusa. Medusa had once been a beautiful woman, who was so boastful of her beauty that she was turned into a Gorgon, a winged monster with snakes for hair and dragon scales for skin. Whoever looked at Medusa's face would turn to stone.

Perseus needed help to accomplish this task. He coerced three nymphs to help him find the Gorgons and to give him the three things he would need to succeed: a pair of winged sandals which allowed him to fly anywhere; a magic helmet which would allow him to see without being seen; and most important he was given a highly polished shield by the Goddess Athena.

After traveling very far he finally found the three Gorgons. They were asleep. He approached them by walking backwards and using his shield as a mirror so he would not look directly at Medusa. He then cut off her head with a sharp sword given to him by Mercury, and placed it in his bag. Immediately after Perseus had killed Medusa, the winged horse Pegasus arose out of her body.

Perseus jumped on the back of Pegasus and they soared away toward home. As he was passing near Ethiopia he heard

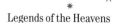

the scream of Andromeda who was about to be attacked by Cetus the sea monster. He immediately turned and noticed the beautiful Andromeda chained to a huge rock. Read how he rescued her in the Cassiopeia legend.

Perseus and Andromeda were happily married except for one sorrowful incident. While participating in a discus-throwing event, he accidently struck a bystander and killed him. The

bystander was his grandfather, King Acrisius, fulfilling the oracle's prophecy that Perseus would cause his grandfather's death. He was so saddened by the tragedy that he gave away the kingdom he inherited.

Legends of the Pleiades

See the legend of Ursa Major. The Pleiades are the seven daughters of Atlas and Pleione. It is said that Orion the hunter tried to kidnap Pleione while she was walking with her daughters. Fortunately, they escaped, but as you may see, as the Pleiades moves across the sky Orion is not far behind.

Among the Aborigines in Central Australia is the legend that the Seven Sisters hungered for some wild figs which could not be found in the sky world, but only on Earth. They therefore came down to Earth. When they arrived they were frightened by the new surroundings and hid in a cave. They were unaware that Nirunja, who lives in Orion, and wanted to make love to the sisters, saw them leave their home in the sky. He secretly followed them down to Earth. When he saw

them entering the cave he decided to wait until nightfall when they would be asleep. He then built a camouflage of fig leaves and slowly crept toward the sleeping beauties. As he snuggled among them they awoke and fought their way to the rear of the cave where they escaped through a small crevice in the rock and flew up to their home in the sky. Nirunja, enraged, ran out of the cave, climbed to the top of the mountain and raced after the sisters. Just as he was about to catch them Taurus the Bull, who lives between the seven sisters and Nirunja's home in Orion, awakened from his sleep and faced Nirunja threatening him with his gigantic horns. Nirunja stared at Taurus, realized he could not get past him, and in frustration returned to his home in the three stars of Orion.

The Maori of New Zealand refer to the Pleiades as Matariki which means 'little eyes', but which also refers to a woman. They visualize its seven visible stars as Matariki and her six daughters. When the Pleiades appears before dawn it is considered the beginning of a new year, at which time the seven women are greeted with songs of hope for the future and songs of tears for the departed. It is a time for festivities and offerings of young shoots of sweet potatoes to Matariki, since she and her daughters watch over and protect their crops.

The Masai of East Africa referred to the Pleiades as the 'rain stars', while the Zulus of South Africa refer to them as the 'digging stars', since they appear at the beginning of the rainy season and denote the time to plow the land.

Legend of Sagitta the Arrow

This small constellation, next to the constellation of Aquila, commemorates the magic arrow of Hercules which was used to kill Aquila, Jupiter's pet eagle, which was inflicting such agony upon Prometheus. See the legend of Aquila.

Legend of Sagittarius

Sagittarius is considered to be a centaur – half horse, half man, a creature with the power of a horse and the understanding of a person.

Among the ancient Greeks the king of the centaurs was Chiron, the kindest and wisest centaur. He was the teacher of Hercules the great hunter, Aesculapius the father of medicine, Achilles, and Jason who sought the golden fleece. It was Chiron who arranged the stars in the order we now see them.

During one of the travels of Hercules he had the opportunity to befriend Pholos, the son of Chiron, when Pholos was in danger. Chiron was very grateful to Hercules for befriending his son when he was in need of help. He therefore placed in the constellation Sagittarius a centaur who was a great archer in order to guard and protect Hercules from Scorpius the scorpion.

If you look at the heavens at night and watch as Orion sets in the west, you will notice Scorpius rising in the east as if it is following the hunter. But Sagittarius, the Archer, follows Scorpius who once attacked Hercules, always ready to attack it if it threatens Hercules.

Legend of Scorpius

The Polynesians of Tahiti tell the story of the boy Pipiri and his sister Rehua. Their mother, who was quite harsh and short tempered, went fishing for the evening meal. It was not until late at night that she caught enough fish for the family dinner. By the time she returned home her husband and children were asleep. They had gone to bed hungry. She awakened her husband and cooked a wonderful meal. He suggested that she awaken the children who had been quite hungry. She refused despite her husband's insistence and put the children's portion away until the following day. Pipiru and Rehua were awakened by the discussion and were severely hurt by their mother's lack of sensitivity to their hunger. They therefore decided to run away from home. Later that night after their parents were asleep they crept out of the house and ran and ran until they reached a tall hill. They climbed the hill and sank to the ground exhausted. They wept because of leaving their home, but they were determined not to return. Their tears flowed until it actually formed a small pool. Meanwhile just before sunrise their mother awakened, saw their bed stained by their tears and immediately awakened her husband to find the children. She saw their tracks barely visible in the early light and followed their small footprints and trail of tears until they reached the top of the hill where the tracks ended. They were looking around, confused for the children were nowhere to be seen until their father looked up and saw Pipiru and Rehua rising to the stars. They decided to follow. When the children saw their parents getting close to them Pipiru asked a giant stag-beetle to help them escape. He placed the children upon his back and with tremendous speed flew up to the stars where they may be seen in the two bright stars in the tail of Scorpius. The giant stag-beetle went to its home in Antares, the bright star in the body of Scorpius. In the South Pacific when parents are unfair or too harsh, children sing a song about Pipiru and Rehua. See the legend of Orion.

Legend of Taurus

The bull was an ancient symbol of worship. He was revered by the Sumerians as the 'Bull of Light' and by the Egyptians as Osiris-Apis, and was the Golden Calf of biblical times. Taurus, the bull, is the symbol of springtime, which is the time for ploughing and planting, but it is also the symbol of love, which seems to blossom during springtime.

There is a Greek legend that the god Zeus fell in love with the beautiful princess Europa, daughter of King Agenor. Europa had been playing at the seashore. Zeus had been watching her and noticed that when she stopped playing she stood at the edge of the sea and wished that she could go far beyond the horizon. He was so enchanted with her loveliness that he transformed himself into a magnificent white bull. He moved close to the princess and lowered his head. The princess immediately knew that he was offering her the opportunity to fulfil her dream. When she looked into his pleading eyes and felt his wave of love, she climbed onto his back. Zeus, in the form of the bull, dashed into the sea and with great speed swam beyond the horizon to the island of Crete. There he changed himself back to his true form, told her of his love and that he was an immortal god. The princess was so overwhelmed by the intensity and sincerity of his love that she accepted him as her lover. The constellation of Taurus symbolizes this love story.

In the constellation of Taurus is the star cluster Hyades. Hyades was given a place in heaven because she had nursed Dionysus, the son of Zeus.

The Pleiades, also a star cluster and part of Taurus, were the seven daughters of Pleione and the giant Atlas. There are many legends about the Pleiades, but read the legend of Ursa Major, which is also about the Pleiades.

Legend of Ursa Major

When the Earth was very young an American Indian wise man sent his seven sons into the forest to learn how to read the wind. They entered the woods and silently walked while listening to every sound of the wind. When night approached they found a place to rest and to sleep. The stars were bright.

During the night the oldest brother was suddenly awakened by a strange sound. The wind was singing. He could not

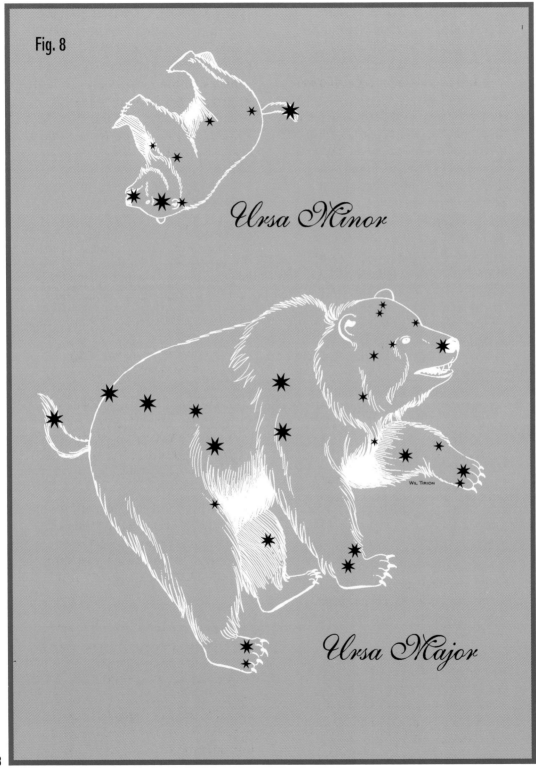

Fig. 8

Ursa Minor

Ursa Major

WIL TIRION

read the wind song, but as he looked to the stars he saw a bright flickering in the Pleiades. He was startled. It appeared to be beckoning for it was flickering in rhythm with the wind song.

He immediately awakened his brothers to listen to the song and to help read the wind. They joined hands and began to dance. The song became stronger and their dance more intense. Suddenly they began to rise toward the flickering star who was the youngest of the seven sisters of the Pleiades. She had fallen in love with the youngest brother Mizar. Since then, Mizar and his love, given to him by the wind song, can be seen by those with sharp eyes in the handle of the Big Dipper – the home of the seven brothers.

This legend has been derived from a Mongolian and an American Indian legend.

Legend of Virgo

Virgo symbolizes the Earth goddess and the goddess of fertility. It is also the symbol of harvest time.

In ancient Babylonia they spoke of the time when the Earth was dark. The plants would not grow and the animals did not give birth. This was the time when Ishtar, the Chaldean goddess of earth and fertility, went through the seven gates of the underworld to find her husband Tammuz, who had been slain by a wild boar and taken to the underworld. As soon as Ishtar entered the first gate to the underworld, the Earth darkened. When she reached the underworld the Queen of Hell refused

to give up Tammuz. When the gods over Earth sent a message to the Queen of the underworld to release Tammuz or be destroyed, Ishtar and Tammuz were sprinkled with magic water, set free and ascended through the seven gates of Hell to Earth. As they reentered Earth spring began, flowers bloomed and the sun warmed the land.

Legends of the Milky Way

The Milky Way has been thought of as the pathway to the home of Zeus/Jupiter. It was also considered the path of Phaeton's wild ride across the sky in the sun chariot.

The Chinese and the Japanese saw it as the silver celestial river.

The Norsemen believed the Milky Way to be the path traveled by the departed souls going to Valhalla. In ancient Wales it was the silver road to the castle of the king of fairies, Caer Groyden.

The Algonquin Indians believed it to be the path of the departed spirits on their way to their villages in the sun. Their path is marked by the stars, which are campfires that guided them along the path.

The Bushmen of the Kalahari Desert in Central Africa speak of the time when a famous hunter was lost in the dangerous bush. Despite days of searching for his village he failed to find the right path. One night, depressed and fatigued, he rested by the edge of a river and prayed to his gods to help him. Hours later, while looking at the sky, he was suddenly aware of a stream of sparkling stars that seemed to point in one direction. He immediately arose and followed that path in the sky and eventually reached his home. His wife, who realized he must have been lost, was throwing embers of the campfire into the sky to form the path that brought him safely back to her.

The Dogons of Africa had a similar legend in that their god Amma threw pellets of earth into the sky thereby forming the Milky Way.

Among some of the natives of the Andes the Milky Way was considered a river on which the spirits of the dead periodically travelled to return to the land of the living in order to perpetuate communication between the living and the dead.

Among Indigenous Australians there exists a secular description of common knowledge about the Milky Way. A married woman had fallen in love with another man. When she became unfaithful to her husband she tried to hide her affair and therefore lied to him to protect herself and her lover, but he was aware of what she had done. He ordered her to build a large fire, When it was very high he grabbed her and threw her into the flames only to see her immediately float up to the sky where she may be seen as a dark patch in the river of stars, the Milky Way.

The Polynesians also speak of Tane, the son of Rangi who was the sky and light, as the god of the forest, beauty and light as well as the god of the fairies.

Rangi and Earth were together in an embrace until Tane separated them, placing the sky high above earth, so that there would be light between them. Tane then threw a basket of stars into the sky to form the Milky Way. Some believe that the Milky Way is the body of Tane's father Rangi.

Some Polynesians consider the Milky Way to be the 'water of life'. They speak of a magnificent blue shark that liked to eat people. It was a pet of the gods. When two young men decided to kill it the gods intervened and placed the shark in the sky where it swims along its river, the Milky Way.

Part 4 There's more to see!

Circumpolar constellations

If an observer is north of latitude 40 degrees, certain constellations may always be seen above the northern horizon circling around Polaris, the Pole Star. These six circumpolar constellations are the Big Dipper (Ursa Major), the Little Dipper (Ursa Minor), Cassiopeia, Cepheus, Draco and Camelopardalis, which is too faint to include (see Fig. 1).

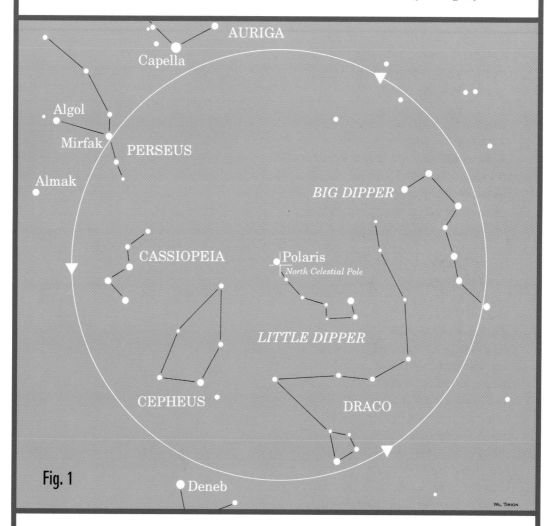

Fig. 1

WIL TIRION

The illusion of these constellations rotating around Polaris is due to Earth's rotation around its own axis. If a line were drawn from the south pole through the center of the earth through the north pole and extended to the celestial sphere it would end approximately 0.8 degrees from Polaris.

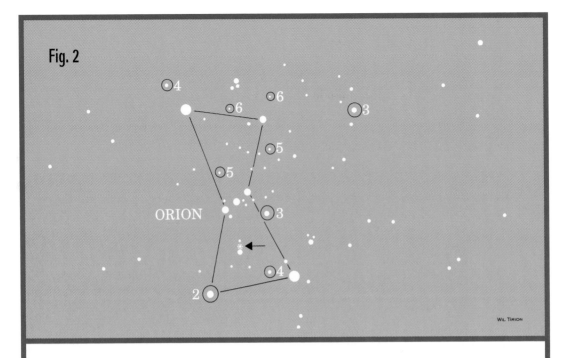

Fig. 2

ORION

WIL TIRION

Test of vision

Although there are 200 billion stars like our Sun in our galaxy, we can only see about 2500 of them at one time above the horizon under ideal conditions.

Color test

Stars vary in color. These colors depend on a star's temperature (just as the color of a flame depends on its temperature). The coolest stars are red, the hottest blue. If you look closely, you can see some of these colors for the brightest stars. Try looking at some of those listed on this page.

Reddish
Antares in Scorpius
Aldebaran in Taurus
Betelgeuse in Orion

Yellowish
Capella in Auriga

White
Sirius in
Canis Major
Formalhaut
in Piscis Austrinis
Altair in Aquila

Blue White
Vega in Lyra
Rigel in Orion
Regulus in Leo
Spica in Virgo
Castor in Gemini
Deneb in Cygnus

Orange
Arcturus in
Bootes

Yellow White
Procyon in
Canis Minor

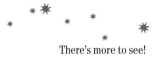

Star brightness test

Most people can see fifth magnitude stars, some can see sixth and even seventh magnitude stars (see page 5). Remember, the higher the magnitude, the dimmer the star. Orion is an excellent testing ground. Look for the hazy nebula (a mass of dust and gas) below Orion's belt (marked with an arrow on Fig. 2). The numbers show the magnitude (brightness) of each star.

Can you see the galaxy in Andromeda (marked by the arrow on Fig. 3)?

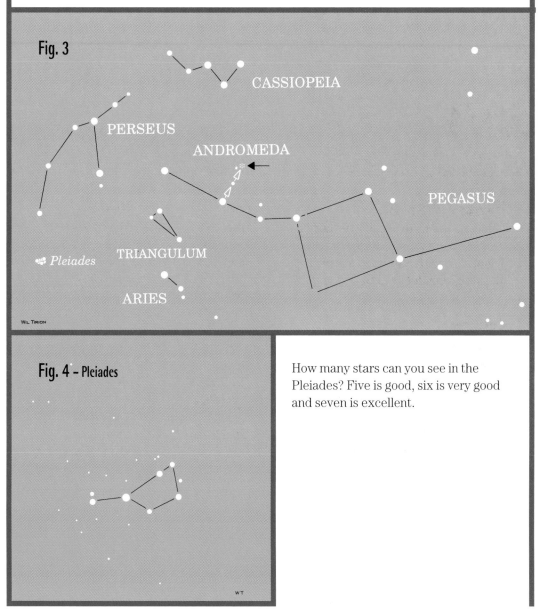

Fig. 3

CASSIOPEIA

PERSEUS

ANDROMEDA

PEGASUS

Pleiades

TRIANGULUM

ARIES

WIL TIRION

Fig. 4 - Pleiades

WT

How many stars can you see in the Pleiades? Five is good, six is very good and seven is excellent.

In the Big Dipper try to see Alcor hidden next to Mizar, Star 2. Alcor has a magnitude of four (see Fig. 2 in Part 2).

Fig. 5 - Little Dipper

2.0
● Polaris

● 4.4

● 4.2

5.0 ● ● 4.3

● 2.1

3.1 ●

The Little Dipper and Bootes are also good tests of vision.

By the way, the twinkling of the stars is not the star itself twinkling, but the effect of atmospheric air currents that break up the rays of light. If the air is very turbulent even the planets may twinkle.

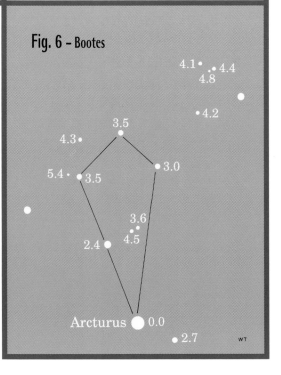

Fig. 6 - Bootes

4.1 ● ● 4.4
4.8
●

● 4.2

4.3 ●   3.5

5.4 ● ● 3.5   ● 3.0

●

3.6

4.5

2.4 ●

Arcturus ● 0.0

● 2.7   WT

Planets

To locate and identify planets

As the planets orbit the Sun they all lie along the same plane as the orbit of Earth except for Pluto. That plane or orbital path is called the ecliptic plane. There are 13 constellations on that same plane. They constitute the Zodiac. However, traditionally the 13th constellation, Ophiuchus, is not included in the zodiacal group.

It is not possible to place the planets into a fixed star map since they are moving in relation to the starry background. To locate and identify planets we must realize that when we see a planet it is always against the background of a different constellation.

Some general points of interest.

1. There are five planets visible to the naked eye: Mercury, Venus, Mars, Jupiter and Saturn.

2. Mars has a distinctive reddish tint.

3. Mercury is difficult to see since it is always seen just after the sun has set and therefore it is against a fairly bright background.

4. The best time to see Venus or Mercury is after sunset in the west or before dawn in the east.

There's more to see!

Planet locations

2004	Jan.	Feb.	Mar.	Apr.	May	Jun.	Jul.	Aug.	Sep.	Oct.	Nov.	Dec.
Venus	CAP	AQR	PSC	TAU	TAU	TAU	TAU	TAU	GEM	LEO	VIR	LIB
Mars	PSC	ARI	ARI	TAU	TAU	GEM	CNC	LEO	LEO	VIR	VIR	LIB
Jupiter	LEO	LEO	LEO	LEO	LEO	LEO	LEO	LEO	VIR	VIR	VIR	VIR
Saturn	GEM	GEM	GEM	GEM	GEM	GEM	GEM	GEM	GEM	GEM	GEM	GEM

2005	Jan.	Feb.	Mar.	Apr.	May	Jun.	Jul.	Aug.	Sep.	Oct.	Nov.	Dec.
Venus	OPH	SGR	AQR	PSC	ARI	TAU	CNC	LEO	VIR	LIB	SGR	SGR
Mars	SCO	SGR	CAP	CAP	AQR	AQR	PSC	PSC	ARI	TAU	ARI	ARI
Jupiter	VIR	VIR	VIR	VIR	VIR	VIR	VIR	VIR	VIR	VIR	VIR	LIB
Saturn	GEM	GEM	GEM	GEM	GEM	GEM	CNC	CNC	CNC	CNC	CNC	CNC

2006	Jan.	Feb.	Mar.	Apr.	May	Jun.	Jul.	Aug.	Sep.	Oct.	Nov.	Dec.
Venus	CAP	SGR	SGR	CAP	PSC	ARI	TAU	GEM	LEO	VIR	LIB	OPH
Mars	ARI	ARI	TAU	TAU	GEM	CNC	CNC	LEO	VIR	VIR	VIR	LIB
Jupiter	LIB	LIB	LIB	LIB	LIB	LIB	LIB	LIB	LIB	LIB	LIB	LIB
Saturn	CNC	CNC	CNC	CNC	CNC	CNC	CNC	CNC	LEO	LEO	LEO	LEO

2007	Jan.	Feb.	Mar.	Apr.	May	Jun.	Jul.	Aug.	Sep.	Oct.	Nov.	Dec.
Venus	SGR	AQR	PSC	ARI	TAU	GEM	LEO	SEX	CNC	LEO	LEO	VIR
Mars	OPH	SGR	CAP	CAP	AQR	PSC	ARI	TAU	TAU	GEM	GEM	GEM
Jupiter	OPH	OPH	OPH	OPH	OPH	OPH	OPH	OPH	OPH	OPH	OPH	OPH
Saturn	LEO	LEO	LEO	LEO	LEO	LEO	LEO	LEO	LEO	LEO	LEO	LEO

2008	Jan.	Feb.	Mar.	Apr.	May	Jun.	Jul.	Aug.	Sep.	Oct.	Nov.	Dec.
Venus	LIB	SGR	CAP	AQR	PSC	TAU	GEM	LEO	VIR	LIB	OPH	SGR
Mars	TAU	TAU	TAU	GEM	GEM	CNC	LEO	LEO	VIR	VIR	LIB	OPH
Jupiter	SGR	SGR	SGR	SGR	SGR	SGR	SGR	SGR	SGR	SGR	SGR	SGR
Saturn	LEO	LEO	LEO	LEO	LEO	LEO	LEO	LEO	LEO	LEO	LEO	LEO

2009	Jan.	Feb.	Mar.	Apr.	May	Jun.	Jul.	Aug.	Sep.	Oct.	Nov.	Dec.
Venus	AQR	PSC	PSC	PSC	PSC	PSC	TAU	ORI	CNC	LEO	VIR	LIB
Mars	SGR	SGR	CAP	AQR	PSC	ARI	ARI	TAU	GEM	GEM	CNC	LEO
Jupiter	SGR	CAP	CAP	CAP	CAP	CAP	CAP	CAP	CAP	CAP	CAP	CAP
Saturn	LEO	LEO	LEO	LEO	LEO	LEO	LEO	LEO	LEO	VIR	VIR	VIR

Positions are for the first day of each month

OPH = Ophiuchus, SGR = Sagittarius, CAP = Capricornus, AQR = Aquarius, PSC = Pisces, ARI = Aries, TAU = Taurus, ORI = Orion, GEM = Gemini, CNC = Cancer, LEO = Leo, SEX = Sextans, VIR = Virgo, LIB = Libra, SCO = Scorpius.

Binocular sights

Binoculars are rated by their magnifying power and by the diameter of the objective lens (the front lens). The ability of binoculars to increase the amount of light seen by the eye depends upon the diameter of the objective lens. The maximum opening of the pupil of the human eye is approximately 8 millimeters. This permits you to see stars with a magnitude of six and possibly seven. Binoculars with a rating of 7 x 32 would have a light-gathering ability nearly 20 times as great as your naked eye since the area of the objective lens is almost 20 x larger than the area of your pupil. This greater light-gathering power enables you to see star clusters, double stars, nebulae and galaxies that you could not see with your naked eye, and single stars with a magnitude of nine or ten.

A star cluster is a group of stars that appear very close together. A double star is actually two stars so close together that they appear as one star. Nebulae are clouds of dust and gas that may appear as dark areas against a background of stars, or with a faint glow if a luminous star is nearby. Galaxies are star systems similar to our own Milky Way. They are sometimes referred to as 'Island Universes'.

There's more to see!

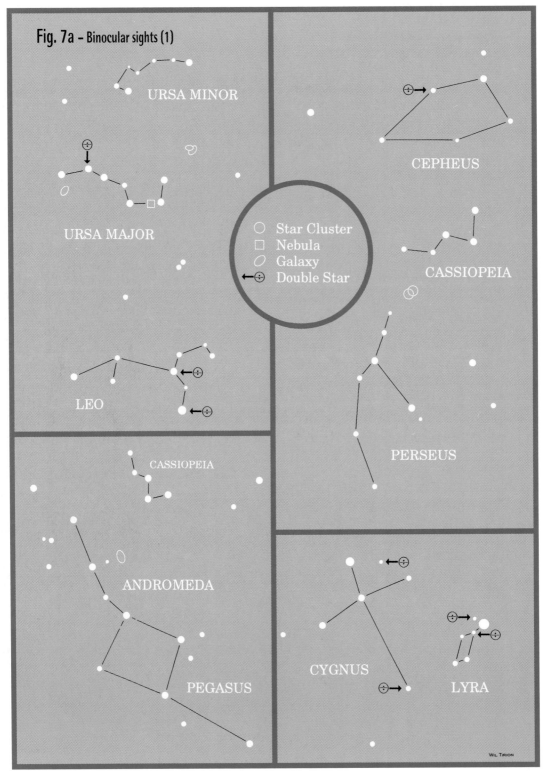

Fig. 7a – Binocular sights (1)

URSA MINOR

URSA MAJOR

○ Star Cluster
□ Nebula
⌀ Galaxy
←⊕ Double Star

LEO

CEPHEUS

CASSIOPEIA

PERSEUS

CASSIOPEIA

ANDROMEDA

PEGASUS

CYGNUS

LYRA

WIL TIRION

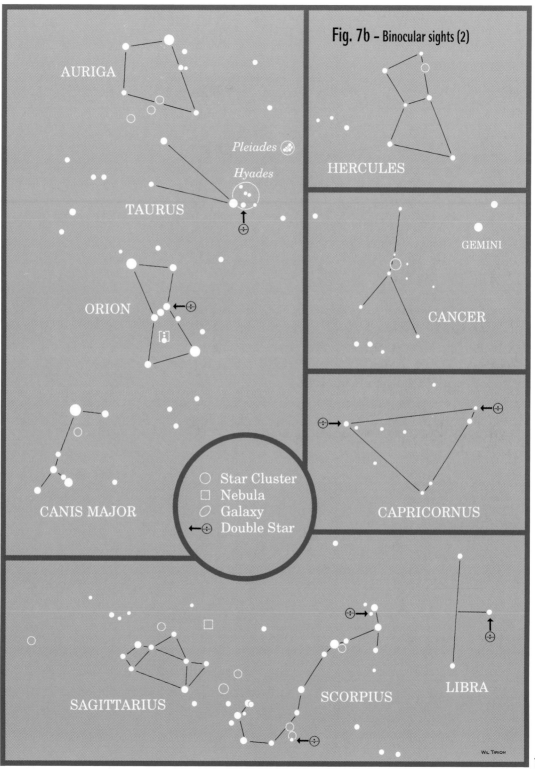

There's more to see!

Fig. 7b – Binocular sights (2)

AURIGA

Pleiades

Hyades

TAURUS

HERCULES

GEMINI

ORION

CANCER

○ Star Cluster
□ Nebula
⊘ Galaxy
←⊕ Double Star

CANIS MAJOR

CAPRICORNUS

LIBRA

SAGITTARIUS

SCORPIUS

WIL TIRION

Meteor showers

As Earth moves in its orbit around the Sun it passes through concentrations of small particles of rock and iron – meteors, the debris of comets or asteroids. Occasionally larger fragments penetrate our atmosphere and strike Earth. These are meteorites. Very rarely a small asteroid (a minor planet) may be large enough to produce a giant crater on Earth. When these objects strike the Earth's atmosphere there is intense friction causing most of the meteors to burst into flame. We see those flashes of flame as 'shooting stars'. They are best seen after midnight.

Shooting stars may be seen almost every night. Occasionally they occur in large numbers producing 'meteor showers'. These showers occur at approximately the same time every year.

The most prominent meteor showers

Name of shower	Date	Constellation location
Quadrantids	Jan. 2–5	Between Bootes and handle of Ursa Major
Eta Aquarids	May 5–6	Northern part of Aquarius
Perseids	Aug. 11–13	Perseus
Orionids	Oct. 20–23	Between Orion and Gemini
Taurids	Nov. 2–5	Between Pleiades and Hyades
Leonids	Nov. 17–18	Near head of Leo
Geminids	Dec. 13–14	Near Castor of Gemini

A massive shower of shooting stars appear every 33 years in the region of Leo. The last period was on November 17–18, 1999. There may be as many as 10 000 meteor flashes per hour.

The brightest stars

	Name	Apparent brightness (Mag.)	Distance in light years
1.	Sun	−26.8	8+ min
2.	Sirius	−1.46	8.7
3.	Canopus	−0.72	1200
4.	A. Centauri	−0.27	4.3
5.	Arcturus	−0.04	36
6.	Vega	0.03	26
7.	Capella	0.08	42
8.	Rigel	0.1	910
9.	Procyon	0.38	11.3
10.	Achernar	0.5	85
11.	B. Centauri	0.6	460
12.	Altair	0.77	16
13.	Betelgeuse	0.4−0.13	310
14.	Aldebaran	0.9	68
15.	A. Crucis	0.9	369
16.	Antares	0.98	330
17.	Spica	1.0	260
18.	Pollux	1.1	36
19.	Fomalhaut	1.2	22
20.	Deneb (Cyg.)	1.3	1600
21.	B. Crucis	1.3	570
22.	Regulus	1.4	85
23.	Adhara	1.5	490
24.	Castor	1.6	45
	Polaris	2.1	700

Navigational stars

Stars have guided navigators as they crossed the oceans and explored foreign lands. Fifty-seven stars have been designated as navigational stars

Star	Constellation
Acamar	Eridanus
Achernar	Eridanus
Acrux	Crux
Adhara	Canis Major
Aldebaran	Taurus
Alioth	Ursa Major
Alkaid	Ursa Major
Al Na'ir	Grus
Alnilam	Orion
Alphard	Hydra
Alphecca	Corona Borealis
Alpheratz	Andromeda-Pegasus
Altair	Aquila
Ankaa	Phoenix
Antares	Scorpius
Arcturus	Bootes
Atria	Triangulum Australe
Avior	Carina
Bellatrix	Orion
Betelgeuse	Orion
Canopus	Carina
Capella	Auriga
Deneb	Cygnus
Denebola	Leo
Diphda	Cetus
Dubhe	Ursa Major
El Nath	Taurus
Eltanin	Draco
Enif	Pegasus
Fomalhaut	Piscis Austrinis

Star	Constellation
Gacrux	Crux
Gianah	Corvus
Hadar	Centaurus
Hamal	Aries
Kaus Australis	Sagittarius
Kochab	Ursa Minor
Markab	Pegasus
Menkar	Cetus
Menkent	Centaurus
Miaplacidus	Carina
Mirfak	Perseus
Nunki	Sagittarius
Peacock	Pavo
Pollux	Gemini
Procyon	Canis Minor
Rasalhague	Ophiuchus
Regulus	Leo
Rigel	Orion
Rigil Kentaurus	Centaurus
Sabik	Ophiuchus
Schedar	Cassiopeia
Shaula	Scorpius
Sirius	Canis Major
Spica	Virgo
Suhail	Vela
Vega	Lyra
Zubenelgenubi	Libra

Minor constellations

Constellations dimmer than 3.5 magnitude

Antlia	<4.3
Apus	<3.8
Caelum	<4.5
Camelopardalis	<4.3
Cancer	<3.9
Coma Berenices	<4.3
Corona Australis	<4.1
Crater	<3.6
Equuleus	<3.9
Fornax	<4.1
Horologium	<3.9
Lacerta	<3.8
Leo Minor	<3.8
Mensa	<5
Microscopium	<4.7
Monocerus	<3.9
Norma	<4
Octans	<3.8
Pisces	<3.6
Pyxis	<3.7
Sculptor	<4.3
Scutum	<3.9
Sextans	<4.5
Volans	<3.8
Vulpecula	<4.4

Constellations Index

Constellations marked with a colored star contain one or more stars that are at least as bright as magnitude 3.

Triangles

DEMCO